T0341319

WATER LOSS ASSESSMENT IN DISTRIBUTION NETWORKS: METHODS, APPLICATIONS AND IMPLICATIONS IN INTERMITTENT SUPPLY

Taha M. AL-Washali

WATER LOSS ASSESSMENT IN DISTRIBUTION NETWORKS: METHODS, APPLICATIONS AND IMPLICATIONS IN INTERMITTENT SUPPLY

DISSERTATION

Submitted in fulfillment of the requirements of
the Board for Doctorates of Delft University of Technology
and
of the Academic Board of the IHE Delft
Institute for Water Education
for
the Degree of DOCTOR
to be defended in public on
Monday, 21 December 2020, at 12:30 pm
in Delft, the Netherlands

by

Taha Mohammed AL-WASHALI
Master of Science in Integrated Water Resources Management
Cologne University of Applied Sciences, Germany
born in Sana'a, Yemen

This dissertation has been approved by the
promotor: Prof.dr. M.D. Kennedy and
copromotor: Dr.ir. S.K. Sharma

Composition of the doctoral committee:
Rector Magnificus TU Delft   Chairman
Rector IHE Delft   Vice-Chairman
Prof.dr. M.D. Kennedy   IHE Delft / TU Delft, promotor
Dr.ir. S.K. Sharma   IHE Delft, copromotor

Independent members:
Prof.dr. Z. Kapelan   TU Delft
Prof.dr. L. Ribbe   Cologne University, Germany
Prof.dr.ir. L.C. Rietveld   TU Delft
Prof.dr. M.L. Al-Eryani   Sana'a University, Yemen
Prof.dr. M.E. McClain   TU Delft / IHE Delft, reserve member

This research was conducted under the auspices of the Graduate School for Socio-Economic and Natural Sciences of the Environment (SENSE)

CRC Press/Balkema is an imprint of the Taylor & Francis Group, an informa business

Published by:
CRC Press/Balkema
Schipholweg 107C, 2316 XC, Leiden, the Netherlands
Pub.NL@taylorandfrancis.com
www.crcpress.com – www.taylorandfrancis.com
ISBN 978-0-367-76655-9

هذا جُهدَ المُقِلّ، اللَّهم تقبَّلَه على عِلّاتِه.

To those who struggle for water supply..

ACKNOWLEDGMENTS

First and foremost, all praise and glory to Almighty Allah who gave me the strength and perseverance to keep going with this research even under the darkest circumstances. My gratitude to Him will always be, to give me the opportunity to work for the benefit of water supply in poor countries. May this effort be useful for them.

Secondly, I here place on record my sincere thanks and gratitude to my supervisors for years of continuous support, especially in the twenty months I spent in Yemen under suffering and airstrikes. I deeply thank Prof. Maria Kennedy for her continuous guidance and support, even at a personal level. The challenges she encouraged me to take were the main lever for my achievement. I am very grateful to Dr Saroj Sharma who was not only a supervisor but a friend and a colleague full of shrewdness and sophistication. He has always been responsive, professional and with an open door. Without his support, this thesis would never be completed. I am indebted to Dr Frank van Steenbergen (MetaMeta Research) for all that he did and continues to do, words do not help to record his due thanks. I will not forget to place here my highest appreciation to Abraham Mehari Haile, Simon Chevalking, and Jolanda Boots, who have undertaken a lot of burdens in this PhD project. I thank with all my heart Allan Lambert (LEAKSSuite, UK), Francisco Arregui (ITA, Spain), Bambos Charalambous (IWA, Cyprus), Malcolm Farley (IWA, UK), Michel Vermersch (IWA, France), and the other colleagues in the IWA water loss specialist group who provided any form of support. I also thank the colleagues in the Water and Environment Center at Sana'a University, beginning with the late Prof. Abdullah Babaqi (PBUH), Prof. Fadhl AL-Nozaily, Dr Adel AL-Washali, Dr Sharafaddin Salah, and Dr Mansour Haidera for their support and trust.

The attribution of Eng Zeyad Shawagfeh, his daily visits, and his persistence in making the fieldwork a success was a decisive factor when expanding the study cases, after the Yemen war. Thanks and appreciation are also presented to each of Eng Ryadh Al-Shaieb, Eng Faris Hubaish, Eng Khaled Qasim, Eng Hani Al-Koli, Eng Mohammad Modaes, Eng Adel Moudhah, and many other engineers and technicians in the corridors of the water utilities in Sana'a, Zarqa and the other utilities covered in this study. The efforts of the master students, whom I cosupervised their Msc research, remain acknowledged. I thank Meidy Mahardani for his choice to repeat the laboratory tests of water meters and float-valves in Indonesia; Mohamed E. Elkhider for his enthusiasm to work together in the collection and analysis of NRW software tools; Tirelo Mphahlele for the analysis of water theft in South Africa; and Robert Lupoja for establishing MNF in Mwanza. My gratitude also goes to Dimitri Solomatine, Nemanja Trifunovic, David Ferras, Mohanad Abunada, Juan Carlos Chacon Hurtado, Jan Teun Visscher, Anique Karsten, Floor Felix, Bianca Wassenaar, Selvi Pransiska, Paula Derkse, Ellen de Kok, Rachelle Dwarka, Wim Glas,

Matt Luna, and Zaki Shubber. A due word of thanks and acknowledgement goes to Nuffic for funding this fellowship.

I also wish to place on record my appreciation to a group of friends; this clip cannot accommodate expressing proper thanks to them; Rob Stockman, Ahmed El-Ghandour, Ahmed Mahmoud Rajab, Mohanad Abunada, Almotasembellah Abushaban, Basel Biadh, Yousef Albuhaisi, Hesham Elmilady, Shaimaa Theol, Nirajan Dhakal, Iosif Scoullos, Md Ataul Gani, Aftab Nazeer, Md. Ruknul Ferdous, Muhammad Nasir Mangal, Shakeel Hayat, Wahib Al-Qubatee, Musaed Aklan, Mohaned Sousi, Shahnoor Hasan, Zahrah Musa, Muhammad Dikman Maheng, Mary Barrios Hernandez, Aries Purwanto, Sebrian Putra, Khalid Hassaballah, Jeewa Thotapitiya Arachchillage, Tesfay Gebretsadkan, Natalia Reyes Tejada, Thaine H. Assumpcao, Maria Luisa Salingay, Vitali Diaz Mercado, Nazanin Moradi, Marmar Ahmed, Abeer Almomani, Meseret Teweldebrihan, Fatima Elfilali and all PhD and IHE community. I do not forget the lines of my love and thanks to all loved ones in Yemen, especially among them Abdul Jabbar Al-Munthary, Bassam Al-Maktari, Faisal Al-Hakeem, Abdulkhaleq Alwan, Ibrahim Al-Sheari, Walid Amer and all the friends.

I record my love, gratitude and appreciation to my family, who surrounded me with warmth, inspiration, and encouragement throughout the past period, my dear father who insisted on my education, and my beloved wife, brothers and sisters, whose love was the fuel to continue this difficult journey. I specifically recognise the support of my brothers in Germany, Abdulatif AL-Washali and Issam AL-Washali. The passion of my daughter, Maryam, and my son, Ahmed, is so deep and beautiful meaning of life, sorry that I was not available as it should be. Finally, I thank everyone who contributed directly or indirectly to the success of this thesis. Thank you all!

Taha AL-Washali

December 2020
Delft, The Netherlands

SUMMARY

Water utilities around the world struggle with the problem of water loss in distribution networks and make continual efforts to control and reduce it. The global annual water loss is estimated at 128 billion cubic metres, causing monetary losses estimated at 40 billion USD every year. Approximately 74% of these losses occur in developing countries. Leakage deteriorates the technical capacity of the network, affects the level of water service, and poses a risk of contamination. Water losses also undermine the economic viability of water services. While leaks increase the cost of energy and treatment, and require an additional quantity of water to replace the leakage (to meet demand), apparent losses reduce the water utility revenues that should be allocated to meet the increasing costs of operation and maintenance. However, the most significant cost due to water losses remains the great waste of water resources, for which the true environmental cost is difficult to estimate, especially in areas experiencing increasing water scarcity.

The first step in dealing with the problem of water loss in distribution networks is to estimate its magnitude and identify its components. Although this basic step affects the understanding of the water utility with regard to the level, nature, and variance of water loss in the network, it remains a complex procedure in the case of intermittent supply. This is due to several factors, the most important of which are: (i) the fluctuation of water losses in the network according to the variation in the volume of water supplied monthly, seasonally, and annually, (ii) the currently available methods of water loss component assessment were originally developed in continuous supply networks with completely different conditions from those of intermittent supply (with water tanks at households), and (iii) the more significant unauthorised consumption in the network. For these reasons, the assessment of the level of water loss and its components in intermittent supply is a more complex process that requires modification of existing methods or developing new methods for water loss component assessment. The objective of this study was to enable water utilities with intermittent supply to assess, on a regular basis, the level and components of water losses in the network; so that a more effective water loss management strategy can be established, monitored, and executed.

As the volume of water losses in intermittent supply varies according to the variation of the supplied water, this study suggests normalising the volume of water losses in the network. Normalisation enables water utilities with intermittent supply to monitor the level of water losses and reveal the progression or regression of water loss management. The study presents the normalisation procedure using two methods. The first method consists of performing regression analysis by correlating the volume of water loss to the system input volume. This method has been proven to be effective in monitoring the status of water losses and determining the extent of progress or decline of water loss management in the network. However, this method is not useful when benchmarking (or comparing) the performance of a water utility to other utilities with different intermittency levels. For comparing and benchmarking, this study discusses another method; the possibility of extending the normalisation approach of real losses to also include the normalisation of the total water loss and apparent losses. The results demonstrate the possibility of using the 'when system is pressurised' (w.s.p.) adjustment to benchmark the water loss level of water utilities with different supply time periods. However, when the supply time is very low (e.g., less than eight hours per day), this method shows questionable results because the normalisation curve of this method is a power curve. This study therefore suggests linearising this curve and considering a linear relationship between water losses and the average supply time. Ultimately, this remains the only method that is currently available for benchmarking water loss performance in networks with intermittent supply.

In addition, this study conducts an in-depth review of the methods of water loss component assessment that were developed in the context of continuous supply, attempting to analyse their applicability towards intermittently operated networks and identifying and addressing their deficiencies. At a system-wide scale, the top-down water balance method is cost-effective and does not require intensive fieldwork. The accuracy of this method therefore depends on the accuracy of the calculation of the apparent losses, which is critical in the case of intermittent supply. This is because estimation of the inaccuracies associated with the water metres of customers requires an analysis of the flow rates of the float valves in the tanks; and because the estimation of the unauthorised consumption in the network is a very complex issue. This study therefore addresses the assessment of apparent losses and discusses ways to improve it.

On the other hand, estimating the leakage volume in a district metered area (DMA) in the network based on analysing the minimum night flow (MNF) is applicable, in principle, to intermittent supply networks, but requires taking into account several considerations. Carrying out MNF analysis in a DMA in the network and inferring the leakage rate based on the analysis of measurements collected over one day is not possible in the case of intermittent supply. This is due to the presence of ground and elevated tanks on the premises, and application of this method requires ensuring that all these tanks are filled with water. This requires transforming the DMA temporarily from an intermittent to a continuous water supply; this is achieved by supplying water to the DMA continuously for several days to ensure that all tanks are full and the minimum inflow readings in the MNF curve begin to repeat themselves. In this case, the 'minimum night flow' would not necessarily occur exclusively at night, but may also occur in the early hours of the day. In the Zarqa water network (Jordan), the 'minimum flow' occurred at 12:15 AM, 04:45 AM and 07:15 AM. This complicates the estimation of customer consumption during the occurrence of the minimum flow, and in some cases this may reduce the accuracy of this method for the case of intermittent supply, or challenge its idea in the first place. This study shows that this method can be only occasionally applied in intermittent supply networks. The regular and systematic application of this method remains impractical as it requires adjusting (or disturbing) the schedule of water distribution in the network to shift the DMA temporarily from intermittent to continuous supply.

If the volume of real losses is estimated, further breaking it down into subcomponents can also be done either at the system-wide or DMA scale, using bursts and background estimates (BABE) analysis. This analysis enables a clear understanding of the factors affecting the volume of real losses and the impact of water utility policies on the volume of real losses. The BABE analysis shows that the volume of water lost from large bursts in the network is much lower than that from small hidden leaks, because they run for a much longer time. Although this method is useful, it has a disadvantage in that it analyses only a small portion of real losses in intermittent supply networks (e.g., 26% of real losses in the Zarqa water network). This is due to the fact that it is an empirical method that has been developed according to data from networks in developed countries under completely different conditions (higher construction quality as well as different policies and technologies of leakage detection different from that of intermittently operated networks in developing countries).

Considering these obstacles, water loss component assessment in intermittent supply is a process marred by high uncertainties, which in turn challenges the effective planning and feasibility of water loss reduction options. Overestimating the real losses exaggerates the economic feasibility of reduction options, whereas underestimating the volume of real losses limits the economic feasibility of reduction interventions. In this regard, uncertainty analysis assists in improving the output of water loss component assessment because it distinctly indicates which input data should be reviewed and improved in order to obtain more reliable results. The aforementioned deficiencies in the assessment of the water loss level and components are reflected, as expected, in the many (free) software tools for water loss management. However, there is a persistent need to highlight two points when developing or updating water loss management software tools: (i) recognising the importance of supply intermittency for expanding the beneficiaries of these tools, and (ii) the importance of addressing apparent losses in these tools.

Focussing on the apparent losses in intermittent supply, this study proposes a practical method for estimating the apparent losses by establishing a water and wastewater balance, using the apparent loss estimation (ALE) equation. The method relies on two routine measurements to assess the apparent losses in the network: (i) measurements of billed consumption and (ii) measurements of WWTP inflow. The results reveal that the parameters involved in this method have a low sensitivity and the accuracy of the WWTP inflow is of greater importance. Installing a metre with a good accuracy (e.g. $\leq \pm 2\%$) to measure the WWTP inflow therefore enables the regular estimation of apparent losses without the need for extensive fieldwork or relying on sensitive assumptions, as is the case in the other methods. However, this method requires, besides the water network, a central sewerage network.

With regard to estimating the customer meter inaccuracy in intermittent supply networks, the study investigates the impact of three different float valves inside a water tank on the water meter accuracy, using laboratory experiments, field measurements, and hydraulic modelling. The flow rates that pass through the water meter correspond to the float valve flow rates, which are lower than the outflow rates from the tank (in the form of customer consumption), owing to the balancing nature of the tank. The study also examines the effect of the degree of water supply intermittency on the performance of the water meter. In general, intermittency has a positive impact on water meter performance.

Conversely, the customer meter inaccuracy would be a critical issue if the water utility is transformed from intermittent to continuous supply with tanks remaining in the network. In this case, the customer meter inaccuracy becomes critical because the tanks remain full most of the time, and the customer consumption slightly affects the water level in the tank, causing a slight opening of the float valve and introducing lower inflows throughout the day. In this case, the customer meter accuracy is greatly reduced, causing a significant increase in apparent losses.

After estimating the total volume of apparent losses in the network and the losses due to the customer meter inaccuracy, the unauthorised consumption in the network can be calculated. Estimating the unauthorised consumption in the network assists in monitoring and managing this important component in intermittent supply networks. When estimating the apparent losses is not possible through the ALE equation, this study suggests a method and a matrix for initial estimates of the unauthorised consumption, based on the number of permanently disconnected connections from the network. This method and the matrix remain more accurate and objective than arbitrary assumptions based merely on data from other networks. By estimating the volume of apparent losses in intermittent supply networks more accurately, the estimation of real losses becomes more accurate. This enhances the economic analysis of options and interventions for water loss reduction, the economic level of leakage, the economic level of apparent losses, and ultimately the economic level of water loss. Finally, the study proposes a guidance framework for the improved assessment of water loss and its components, which enables water utilities to plan, formulate, and monitor effective water loss management strategies in intermittent supply networks.

SAMENVATTING

Waterbedrijven over de hele wereld worstelen met het probleem van waterverlies in distributienetwerken en spannen zich voortdurend in om dit te beheersen en te verminderen. Het wereldwijde jaarlijkse waterverlies wordt geschat op 128 miljard kubieke meter met een geschatte waarde van 40 miljard USD. Zo'n 74% van dit waterverlies vindt plaats in ontwikkelingslanden. Lekkage verslechtert de technische capaciteit van het netwerk, beïnvloedt het niveau van de watervoorziening en vormt een risico voor besmetting van het water. Waterverliezen ondermijnen tevens de economische levensvatbaarheid van waterbedrijven. Het weglekken van water verhoogt de kosten van energie en waterbehandeling omdat er meer water nodig is om de lekkage te vervangen en te voldoen aan de vraag. De schijnbare waterverliezen verminderen de inkomsten van het waterbedrijf die nodig zijn om te voldoen aan de stijgende exploitatie- en onderhoudskosten. De belangrijkste implicatie van waterverliezen is de grote verspilling van watervoorraden, waarvan de werkelijke milieukosten moeilijk in te schatten zijn, vooral in gebieden met toenemende waterschaarste.

De eerste stap bij het aanpakken van het probleem van waterverlies in distributienetwerken is het inschatten van de omvang van het verlies en het identificeren van de componenten. Deze basisstap versterkt het begrip van het waterbedrijf met betrekking tot het niveau, de aard en de variantie van het waterverlies in het netwerk. Deze stap is echter complexer in het geval van intermitterende levering.

Dit is te wijten aan verschillende factoren, waarvan de belangrijkste zijn: (i) De fluctuatie in het waterverlies in het netwerk als gevolg van de variatie in de hoeveelheid water die maandelijks, per seizoen en per jaar wordt geleverd. (ii) De beschikbare meetmethoden zijn ontwikkeld voor continue functionerende waternetwerken, hetgeen totaal andere omstandigheden zijn dan netwerken met intermitterende voeding (met wateropslagtanks in de woningen), en (iii) het grote waterverlies door ongeoorloofd waterverbruik in het waterleveringsnetwerk.

Om deze redenen is de beoordeling van het waterverlies bij systemen met intermitterende watertoevoer een complexer proces dat aanpassing van bestaande of ontwikkeling van nieuwe analyse methoden vereist. Het doel van deze studie is om waterbedrijven met een intermitterende watertoevoer in staat te stellen om op regelmatige basis het niveau van de verschillende waterverliezen in het netwerk te beoordelen; teneinde het mogelijk te maken om een effectievere strategie voor het beheer van waterverliezen op te stellen, te controleren en uit te voeren.

Aangezien het waterverlies bij systemen met intermitterende watertoevoer varieert met de hoeveelheid van het toegevoerde water, propageert deze studie het normaliseren van het waterverlies volume in het netwerk. Door deze normalisatie kunnen waterbedrijven in deze omstandigheid het waterverliesniveau volgen en het effect van beheersmaatregelen zichtbaar maken. De studie presenteert twee normalisatie methoden. De eerste methode bestaat uit het uitvoeren van een regressieanalyse door het waterverliesvolume te correleren aan het invoervolume van het systeem. Deze methode is effectief gebleken bij het monitoren van de waterverliezen en het bepalen van de mate van verandering in het waterverlies door maatregelen in het netwerk.

Voor het benchmarken (vergelijken) van de prestaties van een waterbedrijf met andere bedrijven met verschillende intermittentieniveaus, is een tweede methode nodig die in deze studie wordt besproken. Bij deze methode wordt de normalisatiebenadering van reële verliezen uitgebreid met de normalisatie van het totale waterverlies inclusief de schijnbare verliezen. De resultaten tonen de mogelijkheid aan om de aanpassing 'wanneer systeem onder druk staat' (w.s.p.) te gebruiken om het waterverliesniveau van waterbedrijven met verschillende intermitentie niveaus te benchmarken. Wanneer de leveringstijd echter erg kort is (minder dan acht uur per dag), levert deze methode twijfelachtige resultaten op omdat de normalisatiecurve van deze methode een vermogenscurve is. Hoewel deze methode praktisch en ongecompliceerd is, overschat ze het niveau van schijnbare verliezen in het netwerk, maar in feite is het de enige methode die momenteel beschikbaar is om het waterverlies van verschillende waterbedrijven met intermitterende levering te vergelijken.

De hier gepresenteerde studie voert een diepgaande evaluatie uit van de toepasbaarheid van methoden voor waterverliezen ontwikkeld voor continue opererende waternetwerken, voor de analyse en verbetering van intermitterend werkende netwerken. Op systeembrede schaal is de top-down waterbalansmethode kosteneffectief en vereist geen intensief veldwerk. De nauwkeurigheid van deze methode hangt af van de precisie in de berekening van de schijnbare waterverliezen, wat cruciaal is in het geval van onderbroken toevoer. De reden hiervoor is dat het schatten van de onnauwkeurigheden die verband houden met de watermeters van klanten, een analyse vereist van de stroomsnelheden in de vlotterkleppen in de tanks; daarnaast is ook de schatting van het ongeoorloofde waterverbruik in het netwerk een zeer complexe kwestie. Deze studie behandelt daarom de beoordeling van schijnbare verliezen en bespreekt manieren om deze te verbeteren.

Het schatten van het lekvolume in een districtmetergebied (DMA) (een geisoleerd gedeelte van het netwerk) op basis van het analyseren van de minimale nachtstroom (MNF) is in principe van toepassing op intermitterend werkende watersystemen waarbij wel een aantal overwegingen moeten worden meegenomen. MNF-analyse uitvoeren in een DMA in het netwerk en het lekpercentage afleiden op basis van de analyse van metingen die gedurende één dag zijn verzameld, is bij intermitterende toevoer niet mogelijk. Dit komt door de aanwezigheid van grond- en verhoogde tanks bij gebruikers. De toepassing van deze methode vereist dat al deze tanks gevuld zijn met water. Dit vereist een tijdelijke transformatie van de DMA van een intermitterende naar een continue watervoorziening. Dit wordt bereikt door gedurende meerdere dagen continu water aan de DMA te leveren om ervoor te zorgen dat alle tanks vol zijn en de minimale instroomwaarden in de MNF-curve zich beginnen te herhalen. In dat geval zal de 'minimale nachtstroom' niet noodzakelijk uitsluitend 's nachts plaatsvinden, maar kan deze ook in de vroege uren van de dag plaatsvinden. In het Zarqa-waternetwerk (Jordanië) vond de 'minimumstroom' plaats om 12:15 uur, 04:45 uur en 07:15 uur. Dit bemoeilijkt de schatting van het verbruik van de klant tijdens het optreden van de minimale waterstroom, en in sommige gevallen kan dit de nauwkeurigheid van deze methode verminderen of zelfs twijfel genereren over de toepasbaarheid ervan. Uit dit onderzoek blijkt dat deze methode slechts incidenteel kan worden toegepast in intermitterend werkende voedingsnetten. De regelmatige en systematische toepassing van deze methode blijft onpraktisch, omdat het schema van de waterdistributie in het geisoleerde deel van het netwerk waar de DMA wordt uitgevoerd tijdelijk moet worden omgezet van intermitterende naar continue toevoer.

Na schatting van de werkelijke verliezen kunnen deze verder worden opgesplitst in subcomponenten, hetzij op systeembrede of op DMA-schaal. Dit kan worden gedaan met behulp van de Bursts and Background Estimates (BABE) analyse, waarin Bursts de identificeerbare lekkage betreft en de Backgound Estimates de niet te identificeren lekkage. Deze analyse maakt een duidelijk inzicht mogelijk in de factoren die van invloed zijn op de omvang van de reële verliezen en de impact van het waterleidingsbeleid op de omvang van de reële verliezen. Uit de BABE-analyse blijkt dat het watervolume dat verloren gaat door grote breuken in het netwerk veel lager is dan dat door kleine verborgen lekken, omdat deze laatste vaak veel langer duren. Hoewel deze methode nuttig is, heeft het het nadeel dat het slechts een klein deel van de reële verliezen in intermitterende voedingsnetten analyseert (bijv. 26% van de werkelijke verliezen in het Zarqa-waternetwerk). Dit komt door het feit dat het een empirische methode is die is ontwikkeld op basis van gegevens van netwerken in ontwikkelde landen onder totaal verschillende omstandigheden (hogere constructiekwaliteit en ander beleid en andere technologieën voor lekdetectie) dan die van intermitterend geëxploiteerde netwerken in ontwikkelingslanden.

Gezien deze obstakels is de beoordeling van waterverliescomponenten bij intermitterende toevoer een proces dat gepaard gaat met grote onzekerheden, die op hun beurt een grote uitdaging vormen voor de effectieve planning en haalbaarheid van het verminderen van waterverliezen. Door de werkelijke verliezen te overschatten, wordt de economische haalbaarheid van reductie-opties overdreven, terwijl het onderschatten van de omvang van de werkelijke verliezen de economische haalbaarheid van reductie-interventies beperkt. In dit opzicht helpt onzekerheidsanalyse bij het verbeteren van de resultaten van de beoordeling van waterverliescomponenten, omdat het duidelijk aangeeft welke inputgegevens moeten worden bekeken en verbeterd om meer betrouwbare resultaten te verkrijgen. De eerder genoemde tekortkomingen in de beoordeling van het waterverliesniveau komen, zoals mag worden verwacht, tot uiting in de vele (gratis) softwaretools voor waterverliesmanagement. Er is echter een belangrijke behoefte om twee punten te benadrukken bij het ontwikkelen of aanpassen van software programmas voor waterverliesbeheer: (i) het aanpassen van de programmas voor het gebruik bij intermiterende waterlevering om het aantal gebruikers uit te breiden, en (ii) het goed omgaan met schijnbare waterverliezen in deze programmas.

Deze studie stelt een praktische methode voor om de schijnbare verliezen in watersystemen met intermitterende toevoer te schatten door een water- en afvalwaterbalans vast te stellen met behulp van de schijnbare verliezen formule, de 'formula for Apparent Loss Estimation (ALE)'. De methode is gebaseerd op twee routinemetingen om de schijnbare verliezen in het netwerk te schatten: (i) metingen van het gefactureerde waterverbruik en (ii) metingen van de instroom in de rioolwaterzuivering (RWZI). De resultaten laten zien dat de in deze methode gebruikte parameters een lage gevoeligheid hebben en dat de nauwkeurigheid van de RWZI-instroom van groter belang is. Het installeren van een meter met een goede nauwkeurigheid (bijv. $\leq \pm$ 2%) om de instroom van de RWZI te meten, maakt het mogelijk om de schijnbare verliezen regelmatig te schatten zonder dat uitgebreid veldwerk of gevoelige aannames nodig zijn, zoals het geval is bij de andere methoden. Voor deze methode is echter naast het waternet ook een centraal rioleringsnet nodig.

Deze studie onderzoekt ook de onnauwkeurigheid van de klantmeter in intermitterende waterleidingssystemen. Hierbij wordt de invloed onderzocht van drie verschillende vlotterkleppen in watertanks op de nauwkeurigheid van de watermeter, met behulp van laboratoriumexperimenten, veldmetingen en hydraulische modellen. De stroomsnelheden in de watermeter komen overeen met de stroomsnelheden in de vlotterklep. Deze snelheden zijn lager dan de snelheden van de uitstroom uit de tank doordat de tank de pieken in het watergebruik van de klant afvlakt. De studie onderzoekt ook het effect van de mate van onderbreking van de watervoorziening op de prestaties van de watermeter. Over het algemeen heeft intermittentie een positieve invloed op de prestaties van de watermeter.

Omgekeerd zou de onnauwkeurigheid van de klantmeter een kritieke kwestie worden als het waterbedrijf overgaat van intermitterende naar continue toevoer, terwijl de tanks blijven aangesloten op het netwerk. In dat geval wordt de onnauwkeurigheid van de klantmeter kritiek omdat de tanks het grootste deel van de tijd vol blijven. In die omstandigheid heeft het verbruik van de klant slechts een lichte invloed op het waterpeil in de tank, waardoor de vlotterklep slechts minimaal wordt geopend met het gevolg van een lage instroomsnelheid. Hierdoor wordt de nauwkeurigheid van de klantmeter aanzienlijk verminderd, waardoor de schijnbare verliezen aanzienlijk toenemen.

xvi

Na het schatten van het totale volume aan schijnbare verliezen in het netwerk en de verliezen als gevolg van de onnauwkeurigheid van de klantmeter, kan het ongeautoriseerde verbruik in het netwerk worden berekend. Het schatten van het ongeoorloofde verbruik in het netwerk helpt bij het bewaken en beheren van deze belangrijke component in intermitterend werkende systemen. Wanneer het schatten van de schijnbare verliezen niet mogelijk is via de ALE-vergelijking, stelt deze studie een methode voor om een initiële schatting te maken van het ongeoorloofde water verbruik op basis van het aantal permanent afgesloten wateraansluitingen. Schattingen met behulp van deze methode en de matrix zijn nauwkeuriger en objectiever dan willekeurige aannames die louter zijn gebaseerd op gegevens uit andere netwerken. Door een betere schatting van de schijnbare verliezen in intermitterend werkende waterleidingsystemen wordt de schatting van de werkelijke verliezen nauwkeuriger. Dit verbetert de economische analyse van mogelijkheden en interventies om het waterverlies te verminderen, het economische niveau van lekverlies, het economische niveau van schijnbare verliezen en uiteindelijk het economische niveau van het waterverlies. Tot slot stelt de studie een kader voor om de beoordeling van waterverlies en zijn componenten te verbeteren, waardoor waterbedrijven effectieve strategieën voor het beheer van waterverlies in intermitterend functionerende waternetwerken kunnen plannen, formuleren en monitoren.

CONTENTS

1
INTRODUCTION

1.1 INTRODUCTION

Earth is known as the blue planet because of the abundance of water, however half of the planet's population will be living in water-stressed areas by 2025 (WHO/UNICEF 2017). Only 0.77% of the water on Earth constitutes available freshwater for human use (Shiklomanov 1993). In 2019, water scarcity was ranked as the fourth largest global risk in terms of potential impact (WEF 2019). According to Mekonnen and Hoekstra (2016), two-thirds of the world's population experience severe water scarcity for part of the year and 0.5 billion people face it throughout the year. Water scarcity is exacerbated by population and economic growth as well as the impacts of climate change. If not managed well, water scarcity may lead to disease outbreaks, famine, and conflicts.

Water supply is crucial for a healthy and prosperous life. The collapse of water supply services negatively affects public health, the economy, education, women, human dignity, and triggers disease outbreaks. Water supply is central to social and economic development. According to the WHO/UNICEF Joint Monitoring Program, 2.2 billion people do not have safely managed drinking water services, 785 million people lack basic water services, 144 million people drink untreated surface water, and 2 billion people use a contaminated drinking water source, causing 485,000 diarrhoeal deaths each year (WHO/UNICEF 2017). Around 1,000 children die daily due to preventable water- and sanitation-related diarrhoeal diseases. These facts were the driving force for the water and sanitation sustainable development goal aiming for universal and equitable access to safe and affordable drinking water by 2030 (UN 2015). Achieving this goal is only possible if water utilities can provide cost-effective water services. However, according to the International Benchmarking Network (IBNET), 37% of water service providers worldwide cannot cover their basic operation and maintenance (O&M) costs. In low- and middle-income countries, 70% of water utilities are not able to cover their O&M costs (Danilenko et al. 2014). The crux of enhanced water supply efficiency lies in reducing water and revenue losses in distribution networks.

1.2 GLOBAL LEVEL OF NON-REVENUE WATER

Non-revenue water (NRW) remains one of the most pressing deficiencies for water utilities worldwide (Danilenko et al. 2014; Kingdom et al. 2006). NRW represents water that is supplied but not sold to customers or used by or through the water utility. There exists no water distribution system with 0% losses as all water distribution networks leak, but to different extents. Liemberger and Wyatt (2018) estimated the global level of NRW based on available NRW percentages in the database of the International Benchmarking Network (IBNET 2020), along with global population estimates, country-based per capita consumption data, and population (%) with piped water data based on the WHO/UNICEF Joint Monitoring Program (WHO/UNICEF 2017). These authors estimated the 2018 global volume and cost of NRW at 126 billion m³/yr and US$ 39 billion per year, respectively. This figure is in fact underestimated, because a lower percentage of population with piped water (54%) was used (WHO 2020; WHO/UNICEF 2017). From analysing the global data published by Liemberger and Wyatt (2018), the estimated global NRW volume and cost in 2020 is 128 billion m³/yr and US$ 40 billion per year, respectively. Approximately 74% of these losses occur in developing countries. With a business as usual scenario, Figure 1.1a shows the expected global volume and cost of NRW by 2050, which will be 176 billion m³/yr and US$ 55 billion per year, respectively. Figure 1.1b shows the range of NRW levels worldwide. Most of the global NRW data as a percentage of supplied water lie between 20% and 40%, with an average of 32%.

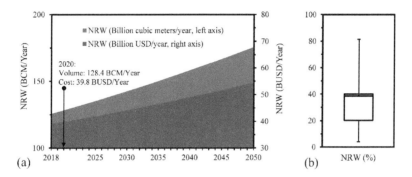

Figure 1.1. Global level of NRW: (a) volume and cost of NRW; and (b) global range of NRW as a percentage of supplied water. Figure developed based on IBNET data from Liemberger and Wyatt (2018)

The impact of NRW is substantial. Leaks affect the technical stability of the water supply system, the operational age of the network, the water quality, and the quality of the water service. From an environmental perspective, leakages should always be addressed to alleviate pressure on water resources as they cause considerable water wastage. If leakage is not addressed, water awareness and conservation campaigns in the municipal sector are rendered insignificant, as the quantity of water conserved through rationing water use inside premises is not as significant as the water that leaks from distribution networks. From an economic point of view, NRW undermines the economic feasibility of the water service. While leaks increase operating costs (treatment and transportation) and require a larger investment, apparent (commercial) losses significantly reduce the utility revenues. The opportunity cost of NRW is the extension of water services to cover new populations, as more than 785 million people still lack access to a basic water service. The level of daily leakage in the world, as of 2006, could serve a further 200 million people (Kingdom et al. 2006), and reducing commercial losses could generate revenues to cover parts of the required capital investment.

Despite its negative impacts, NRW is still high in many countries. Figure 1.2 shows the NRW percentages of different countries around the world, developed based on IBNET and WHO/UNICEF data published by Liemberger and Wyatt (2018). High NRW percentages are predominant in Africa, Asia, and South America. However, expressing the level of NRW as a percentage of supplied water is strongly criticised in NRW practitioners' circles because it is influenced by water consumption levels. Figure 1.3 shows the NRW level in litres per capita per day, which is informative in regard to countries where the NRW problem is more pressing, as is the case in countries of the Arabian Gulf, where significant proportions of the supplied water consist of desalinated sea and brackish water. Figure 1.4 shows the cost of NRW for each country, which is a direct indicator of revenue losses due to the NRW problem.

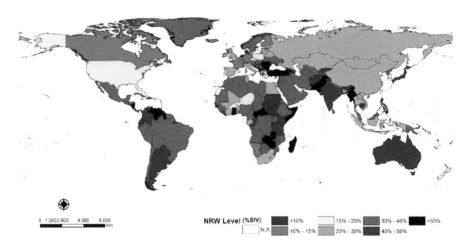

Figure 1.2. World map of NRW as a percentage of supplied water. Map developed based on IBNET data from Liemberger and Wyatt (2018)

1.3 NRW MANAGEMENT

NRW reduction is achieved by reducing leakages and apparent losses in water distribution networks. While apparent losses have been relatively overlooked, significant progress in leakage reduction has been achieved in terms of research and technology. Partitioning the network into several district metered areas (DMA) yields significant benefits, including pressure management, leakage monitoring, leakage detection, maintaining water quality, and asset management.

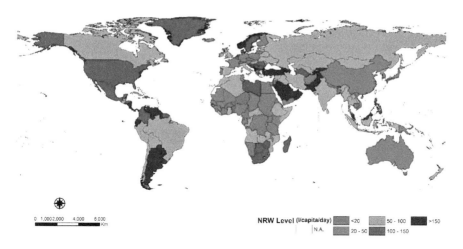

Figure 1.3. World map of NRW in litres per capita per day. Map developed based on IBNET data from Liemberger and Wyatt (2018)

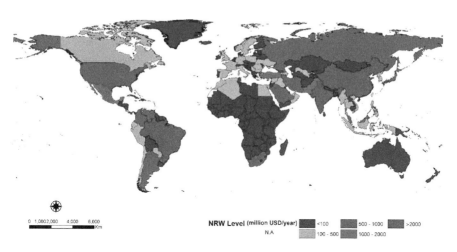

Figure 1.4. World map of NRW cost (million USD/yr). Map developed based on IBNET data from Liemberger and Wyatt (2018)

Several methods exist for network partitioning based on different criteria, including topology, reachability, connectivity, redundancy, and network vulnerability (Deuerlein 2008; Di Nardo et al. 2013b; Galdiero et al. 2015; Morrison et al. 2007). Incorporating the establishment of DMAs with pressure management is fruitful (Alonso et al. 2000; Creaco and Pezzinga 2014; De Paola et al. 2014). Pressure management is the most effective tool for leakage reduction. The use of pressure-reducing valves is a key

5

component of of pressure management (Dai and Li 2014; Vicente et al. 2016). Active leakage detection applies methods to detect, locate, and pinpoint leaks, such as correlating noise loggers (Li et al. 2014; Puust et al. 2010; Wu and Liu 2017). Lambert and Fantozzi (2005) suggested designing the frequency of leakage detection surveys based on economic data, which relies on estimating the rate of rise of leakage (Lambert and Lalonde 2005). Leakage can also be controlled by minimising the repair-response time and asset management (Christodoulou et al. 2008; Creaco and Pezzinga 2014). Figure 1.5 shows the four basic techniques for effective leakage reduction. However, leakage cannot be totally eliminated. Leakage will always occur even in well- and newly-established networks. For this reason, the unavoidable leakage concept has been proposed (Lambert et al. 2014; Lambert and McKenzie 2002). Water utilities should apply the four leakage management techniques to squeeze the current level of leakage shown in Figure 1.5 to an economic level, after which further reducing the leakage becomes uneconomic. From a mere economic perspective, the economic level of leakage (ELL) is reached when the cost to further reduce leakage exceeds the expected benefits, as illustrated in Figure 1.6. (Ashton and Hope 2001; Kanakoudis et al. 2012; Pearson and Trow 2005). There are, however, considerable environmental benefits of leakage reduction that are not recognised by the ELL methodology.

Similarly, Figure 1.6 illustrates the basic techniques for reducing apparent losses, which consist of the water that is consumed by customers but not paid for. Customer meters are prone to under-registration; therefore, the first technique for countering this is effective customer meter management to minimise metering errors in the network.

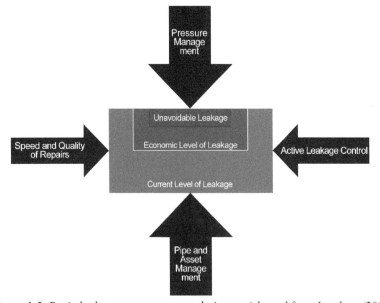

Figure 1.5. Basic leakage management techniques. Adapted from Lambert (2002)

Analysing the water consumption pattern as a basis for appropriate meter selection, accuracy evaluation, and sizing is crucial in water meter management (AL-Washali et al. 2020a; Arregui et al. 2006a; Mutikanga 2014; Van Zyl 2011). Furthermore, large customers (3–4% of customers) utilise more than 50% of the total consumption; therefore, inspecting large customers' meters is vital (Vermersch et al. 2016). The proper sizing and installation of meters as well as an optimal meter replacement policy, are central for the management of meters used by various customers (Arregui et al. 2006a; Van Zyl 2011).

Installing an accurate meter fleet is pointless if the consumption data are not read and processed reliably. Apparent loss management therefore also involves minimising errors in the data acquisition process and reducing misestimates of unmetered consumption (Vermersch et al. 2016). Finally, reducing unauthorised consumption from network components can be achieved by detecting and inspecting illegal connections, bypasses, and water theft (AWWA 2016; Thornton et al. 2008; UNHSP 2012). Customer management, community participation, awareness and communication policies, as well as customer surveys are essential in the control of unauthorised consumption (Al-Washali 2011; Carteado and Vermersch 2010; Farley et al. 2008; Mutikanga et al. 2011a). Water utilities should apply the four techniques of apparent loss management shown in Figure 1.6 to squeeze the current level of apparent losses to an economic level.

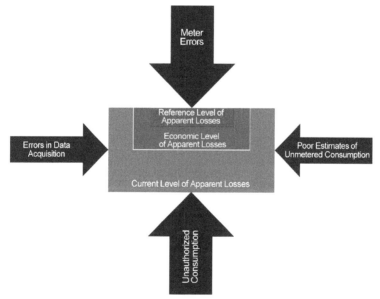

Figure 1.6. Basic apparent loss management techniques. Adapted from Vermersch et al. (2016)

The reference level of apparent losses in Figure 1.6 is set to identify the limit of unavoidable apparent losses that cannot be eliminated from the system. New meters, for instance, have limitations at low flows, and these limitations cannot be removed with the available technology on the market.

Management involves setting targets and utilising available resources to achieve them. Farley and Liemberger (2005) proposed a diagnostic approach to develop a NRW management strategy based on basic questions and outlined typical measures to answer each question, as shown in Figure 1.7. The baseline assessment of NRW components is clearly critical for setting NRW management targets. Once the NRW strategy is defined based on the proposed framework in Figure 1.7, regular monitoring of NRW management progress is key. The most common NRW performance indicator (PI) is the NRW as a percentage of the system input volume, which is difficult to compare internationally (Lambert et al. 2014; McKenzie and Lambert 2004). For example, for a leakage level of 100 l/service connection/day, the water loss percentage varies from 17% for systems with low water consumption to 1% for systems with high water consumption (Lambert and Taylor 2010). For this reason, different NRW PIs have been proposed for target setting as well as comparison and benchmarking (Alegre et al. 2006; Alegre et al. 2016).

Finally, minimising NRW to zero is not technically possible or economically viable. The economic level of NRW should be estimated based on the economic levels of leakage and apparent losses, to identify how NRW management strategies can be most cost-effectively achieved, and to determine the priority with which the components should be tackeled. Farley et al. (2008) suggested identifying the economic level of NRW based on the principle of cost-benefit analysis by comparing the cost of water loss with the cost of undertaking reduction activities, as shown in Figure 1.8. Any further reduction of NRW beyond the economic level is not considered economically feasible.

Question	Task
1. How much water is being lost? – Measure components	Water balance – Improved estimation/measurement techniques – Meter calibration policy – Meter checks – Identify improvements to recording procedures
2. Where is it being lost from? – Quantify leakage – Quantify apparent losses	Network audit – Leakage studies (reservoirs, transmission mains, distribution network) – Operational/customer investigations
3. Why is it being lost? – Conduct network and operational audit	Review of network operating practices – Investigate: historical reasons poor practices quality management procedures poor materials/infrastructure local/political influences cultural/social/financial factors
4. How to improve performance? – Design a strategy and action plans	Upgrading and strategy development – Update records systems – Introduce zoning – Introduce leakage monitoring – Address causes of apparent losses – Initiate leak detection/repair policy – design short-medium-long-term action plans
5. How to maintain the strategy?	Policy change, training and O&M Training: improve awareness increase motivation transfer skills introduce best practice/technology O&M: Community involvement Water conservation and demand management programmes Action plan recommendations O&M procedures

Figure 1.7. Tools for NRW strategy. Source: Farley and Liemberger (2005)

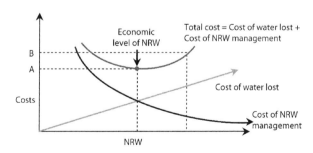

Figure 1.8. Concept of the economic level of NRW. Source: Farley et al. (2008)

1.4 THE NEED FOR RESEARCH

Many water utilities in developing countries attempt to combat the water loss problem with an "indiscriminate shelling" approach, with all the associated field challenges and cognitive bias. A common approach is to either do whatever is possible or just overlook this complex buried problem. In this context, the Zarqa water utility (Jordan) is a good example. This utility established a meter replacement policy where customer meters are replaced when a meter reading reaches 2,500 m^3. The utility established a water loss unit that searches for, among other activities, domestic illegal users in the network. The utility also pushed to implement a megaproject to replace the pipelines for most of the Zarqa network. Together with donors, the utility further equipped and trained leak detection teams to perform frequent leak detection campaigns in the network. To minimise the response time for reported bursts, a complaint reporting system was established to record and track the procedures from the moment a burst is reported until the repair time. Another effective meter reading system was established to record the meter readers' routes, mobilise them regularly, and equip them with hand-held devices to prevent under-estimation of the customers' water consumption. All these measures were adopted several years prior to 2014, however the NRW level in 2014 was as high as 65% of the supplied water. Clearly, "useful interventions" will remain "inadequate blind actions" without proper diagnosis of the water loss level and components. After the establishment of the IWA water balance in Zarqa, the reduction of unauthorised consumption (used for irrigated agriculture in the city's suburbs) and pressure management were found to be the two most promising options for reducing NRW in the Zarqa network.

Assessment of overall loss level in intermittent supply

When losses in the network are problematic, the first step is then to know how much is the level, volume and components of water losses in the network. This is a basic step influencing all subsequent planning and intervention measures to reduce water losses in the network. This critical step is, however, rather difficult in systems with intermittent supply. The overall level of water loss is variable in intermittently operated networks owing to the fluctuation of the water supplied to the network. The water supplied to customers tends to decrease in water-scarce basins (misleadingly suggesting constant loss reduction) and typically increases when alternate water resources exist (misleadingly suggesting greater losses and a worse performance). In an intermittent supply regime, the figures and indicators do not necessarily represent the ground situation. Water loss management practitioners emphasise that expressing water loss as a percentage of the supplied water should be substituted by 'volume-based' indicators (Alegre et al. 2016; Lambert et al. 2014). This solution is, however, not sufficient for monitoring and tracking water losses in an intermittent supply, because the volume of water losses varies with the varying water production. The question arises, what should it be monitored? Water loss volume or water loss status in the network? Although this is a basic problem that affects intermittent supplies worldwide, it has not been sufficiently or explicitly discussed.

Water loss component assessment in intermittent supply

Even if the water loss level is well-monitored and can be reasonably estimated on a regular basis, addressing it as a black box is clearly inefficacious. Breaking down the water loss level into leakage and apparent losses is also critical and complex in a network affected by intermittency. In an intermittent supply regime, the supply scheme is often adjusted by the customers, installing water tanks (typically with attached float valves) on the premises. Water is not available 24/7 in the network, and hence customers use water from the tanks to buffer the discontinuous supply from the network. Given the complexity of networks with intermittent supply, assessing water loss components in such networks using methods that were basically developed for continuous supply can be clearly problematic. Networks with intermittent supplies are associated with many challenges that vary from one country to another, but include some or all of the following: (i) a high level of water loss; (ii) insufficient water resources; (iii) poor governance; (iv) poorly designed and poorly constructed networks; (v) interlinked or multi-fed networks; (vi) data incompleteness; (vii) poor data quality; (viii) a high level of unauthorised consumption; (ix) a lack of technical capacity; and (x) a lack of equipment.

Top-down water balance

In principle, the IWA top-down water balance methodology requires prior estimates of unauthorised consumption and meter inaccuracy in the network. Owing to the complex and hidden nature of unauthorised consumption, it is typically assumed to be at a low level, which can be justifiable in developed countries. However, in developing countries, unauthorised consumption is too critical to assume and too complex to estimate. Arbitrary assumptions of unauthorised consumption affect the estimated volume of leakage and the economic feasibility of leakage reduction interventions. In addition, estimating the customer meter inaccuracy in a network affected by intermittency requires rethinking the adopted approach. Water meters tend to have a low accuracy at low flow rates and good precision at higher flow rates (Arregui et al. 2006a; ISO 2014a; OIML 2013a). The meter accuracy can typically be estimated based on analysing the customers' consumption flow profile and recognising the meter accuracy at each flow (AWWA 2016; ISO 2014b; OIML 2013b), with an eye on the critical low flows. In the case of intermittent supply, the float valve attached to the water tank introduces longer inflows that are lower than the consumption flows, especially when the tank is almost full, substantially affecting meter accuracy. Estimating the meter accuracy based on analysing the consumption flow profile is therefore not reflective of the actual situation, and this approach needs to be adapted to the intermittent supply situation. In light of the above, the accuracy of the IWA top-down water balance methodology in intermittent supply remains questionable. It should be noted that this is the only common methodology to establish a system-wide water balance and the most common method for leakage estimation. This methodology, however, originated in developed countries to estimate leakage volume, playing down the apparent losses in the network, which clashes with the reality in developing countries.

Minimum night flow analysis

Estimating the leakage volume in a small part of the network may also be accomplished by analysing the minimum night flow in a district metered area (DMA). This approach is typically carried out at night between 2:00 AM and 4:00 AM when most customers are sleeping and the inflow into the DMA is predominantly leakage. If there are sufficient representative DMAs in the network, this can provide a reasonable estimate of the network's overall leakage. Nevertheless, applying this approach in intermittent supplies is countered by the challenge of the water tanks in the network. Even when customers are sleeping in a DMA, water will keep flowing into water tanks on their premises between 2:00 and 4:00 AM. This approach is only feasible if all the tanks in the DMA are completely full, which is a challenging task in intermittent supplies given the regular rationing and scheduling of water supply in the network. Even if the leakage can be estimated in a DMA, generalising its leakage level to the entire network is rather uncertain because each DMA differs in terms of size, pressure, and pipe conditions.

Research problem

Estimating the water loss level and components in an intermittent supply regime remains a complex process. The available methods in the literature were developed for continuous supply; they either require adaptation or new methods should be developed to recognise the specific conditions of intermittent supply. Fluctuations in water production, a high level of unauthorised consumption, discontinuous water supply, and an adjusted supply scheme with water tanks and float valves are key issues that should be considered when attempting to analyse the losses in intermittent supply networks. If these are not considered, unmethodical leakage reduction planning and ineffective water loss management are expected, as discussed above for the case of the Zarqa water network.

Research questions

In order to address the research problem discussed in Section 1.4, the following research questions were formulated for this study:

- What are the available methods and tools for assessment of water loss level and components?

- What are the potential, limitations and implications of the application of different water loss component assessment methods in intermittently operated networks?

- How should the water loss level be monitored under intermittent and variant water supply?

- How can apparent losses be assessed in networks under intermittent supply? Namely:

    - How can the volume of apparent losses be estimated in intermittent supply?

    - What is the impact of the float valve and the water tank at customer premises on the accuracy of customer water meters?

    - What are the possible methods for the estimation of unauthorised consumption in water distribution networks?

1.5 OBJECTIVES OF THE STUDY

The ultimate goal of this study was to enable water utilities to manage leakage and apparent losses in intermittent water supply systems. The level, volume, and components of water loss in the network should therefore be assessed, on a regular basis, so that a more effective water loss management strategy can be set, monitored, and fulfilled. To achieve this, the specific objectives of this study were as follows:

1. To review, with an intermittent supply lens, the available methods (Chapter 2) and tools (Chapter 3) for estimating water losses in distribution networks;

2. To formulate an approach to monitor the overall water loss level and water loss performance indicators in intermittent supply networks (Chapter 4);

3. To gain in-depth insight and understanding of the potential, limitations, and implications of water loss component assessment through minimum night flow analysis (Chapter 5), another potential approach (Chapter 6), and the top-down water balance method, under intermittent supply conditions (Chapter 7);

4. To analyse the impact of the water tank equipped with a float valve on the customer's meter accuracy and the level of revenue losses (Chapter 8);

5. To develop methods for water loss component assessment, addressing the apparent losses (Chapter 6) and recognising the high level of unauthorised consumption in developing countries (Chapter 9).

1.6 STRUCTURE OF THE THESIS

This thesis is composed of ten chapters. Chapter 1 presents the background and objectives of the study.

Chapter 2 comprises an up-to-date review of water loss assessment methods in water distribution networks, where the methods and their advantages and limitations are discussed in detail.

An overview of freely available software tools for water loss assessment is provided in Chapter 3. This chapter describes the tools, presents their theoretical background and key features, and discusses their applicability in intermittent supply. Fit-for-purpose guidance on the use of the tools as well as future prospects are also presented in this chapter.

Chapter 4 investigates the influence of the amount of water supplied to a distribution network on the reported level of NRW and proposes two approaches to normalise the level of NRW for target monitoring and benchmarking, using the water network of Sana'a (Yemen) as a case study.

Chapter 5 examines the applicability of minimum night flow analysis in an intermittently operated DMA in the Zarqa water network, Jordan. The impact of generalising the leakage level in a DMA to the entire network is discussed, and its implications on analysing the economic benefits of leakage reduction options are analysed in this chapter.

In Chapter 6, a new method for water loss component assessment is proposed, which may be more applicable in networks with an intermittent supply and a high level of unauthorised consumption. This chapter presents the method development as well as the method application to the case of Sana'a water network.

Chapter 7 analyses the application and accuracy of the different water loss assessment methods in three intermittently operated networks in three developing countries (Jordan, Yemen, and Tanzania). A comparative uncertainty analysis of the different methods is conducted, and the sensitivity of water loss component assessment to leakage reduction planning is discussed. Recommendations on the use of the methods are also presented in this chapter.

The following two chapters address apparent losses. The impact of water tanks with float-valves on the accuracy of customers' water meters under intermittent and continuous supply conditions is investigated in Chapter 8. The influence of different types of float valve and tank sizes on water meter performance is analysed, and the impact of transforming from intermittent to continuous supply in networks with water tanks and float valves is discussed in this chapter.

Chapter 9 deals in detail with the problem of estimating unauthorised consumption. The chapter proposes several methods for its estimation, introducing the underlying concepts behind these methods and demonstrating their application in six case studies in Asia and Africa. Recognising its complexity, this chapter proposes a matrix for initial unauthorised consumption estimation that utilises the number of permanently disconnected customers as an indicator of unauthorised consumption in the network.

Finally, key conclusions of the different chapters are summarised and the future outlook and recommendations for water loss assessment in intermittent supply are presented in Chapter 10.

2

METHODS OF WATER LOSS COMPONENT ASSESSMENT: A CRITICAL REVIEW

This chapter reviews water loss assessment methods in water supply systems. There are three main methods used to assess water loss: Minimum Night Flow (MNF) analysis, Bursts And Background Estimates (BABE), and Top-Down Water Balance. MNF analysis provides actual measurements and requires intensive field work. The limitation of the MNF method is the sensitivity of two parameters: average pressure, which is rarely accurate, and estimation of the night consumption. Assessment of real losses with the factors generated by the BABE model should not be conducted unless there is no other option, owing to its excessive assumptions. Instead, the method should be a supplementary tool to break down the volume of real losses into its sub-components. The Top-Down Water Balance is neither a pressure-dependent nor an extensive-field-work requiring method. However, its assumptions of apparent losses are not appropriate for all utilities. The lack of an objective methodology for estimating unauthorised consumption is a major limitation; consequently, research on its estimation is demanding.

This chapter has been published as: AL-Washali, T., Sharma, S., and Kennedy, M. "Methods of Assessment of Water Losses in Water Supply Systems: a Review." Water Resources Management, 30(14), 4985-5001, 2016.

2.1 INTRODUCTION

Addressing WL is not a one-step action but rather a continuous process. A comprehensive overview of WL management enables the categorisation of its activities into three primary repetitive stages (Figure 2.1): (1) assessment and monitoring of WL, (2) strategisation and planning of cost-effective measures, and (3) implementation of reactive and proactive actions, such as leakage and illegal-use detection.

WL has two main components—real losses (RL) and apparent losses (AL) (Lambert and Hirner 2000). Real losses, also called physical losses, refer to the water that leaks out of pipes and other bursts and leakages. Apparent losses refer to the commercial losses that are not physically lost but rather represent unpaid use by the customer. The apparent losses include illegal water use, customer meter under-registration, and data handling or billing errors. The difference between non-revenue water (NRW) and WL is the amount of authorised consumption that is used legally but not billed or paid for, that is, unbilled authorised consumption (UAC).

WL assessment involves the quantification of WL in a particular system, without considering where the losses are actually taking place (Puust et al. 2010). During the 1990s, WL assessment was more a "guesstimation" process than meticulous science (Liemberger and Farley 2004). However, more recently, WL assessment and management have progressed significantly. Large efforts have been made by the International Water Association (IWA) and other organisations to promote new concepts and methods for improving WL management (Vermersch and Rizzo 2008).

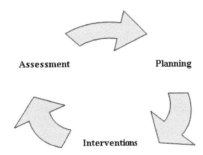

Assessment   Planning

Interventions

Figure 2.1. Water loss management stages

This chapter focuses on the assessment of WL, quantifying its components and addressing the question of how much water is being lost in each component. The chapter summarises the current state of knowledge and critically reviews the main methods of WL assessment. The limitations and potential use and improvement for each method are also highlighted.

2.2 STANDARD TERMINOLOGY

WL is a common challenge for water utilities throughout the world. To share experiences and lessons learned by various water utilities, there should be a common language. Until the early 1990s, there had been no standard terminology for WL components or quantification (Frauendorfer and Liemberger 2010). Instead, international comparisons encountered a wide diversity of formats and definitions for WL components and sub-components (Alegre et al. 2000; Lambert 2002). There was avoidable misunderstanding by individual countries for describing and calculating WL (Lambert and Hirner 2000). An example of such misinterpretation is the use of the term 'unaccounted for water' (UFW), which had several interpretations worldwide and no generally accepted definition. Consequently, IWA recommended not using the term UFW (Frauendorfer and Liemberger 2010; Lambert 2002; Lambert and Hirner 2000). Instead, there was a need for standard terminology as a precondition for calculating internationally comparable WL balances, experiences, and performance indicators (Alegre et al. 2000; Farley and Trow 2003; Frauendorfer and Liemberger 2010; Lambert 2002; Lambert et al. 1999; Lambert and Hirner 2000).

Over the last 20 years, IWA and other organisations have developed tools and methodologies to help utilities evaluate and manage WL in an effective manner (EPA 2013; Frauendorfer and Liemberger 2010). The IWA Water Loss Task Force has developed an international standard water balance with clear definitions, as presented in Figure 2.2 (Farley and Trow 2003; Lambert et al. 2014). The IWA approach was first introduced in an IWA international report by Lambert and Hirner (2000) and quickly gained wide acceptance and promotion by many national and international organisations, including the American Water Works Association (AWWA), US Environmental Protection Agency (EPA), Asian Development Bank, and the World Bank (EPA 2010; Lambert et al. 2014; Radivojević et al. 2008). The IWA terminology which was first presented in Lambert and Hirner (2000) has been revised over time, fine-tuned, and elaborated with minor modifications and highlights, as in Lambert et al. (2014) and Vermersch et al. (2016).

The standardisation of terminology should be considered as a significant achievement in the field. It allows valuable and more precise comparisons and discussions among water utilities and WL researchers and specialists. It should be adopted to support a common communication language among specialists. However, there is still a room for discussion. The wording of some terms can be revised. The term "apparent losses" seems inappropriate because apparent losses are not apparent. The term claims clarity and simplicity in its sub-components, while in fact, it is not the case, especially in developing countries where hidden unauthorised consumption is significant. Referring to apparent losses, the World Bank uses the term "commercial losses," which offers an indication of the nature of its sub-components.

System Input Volume	Authorized Consumption	Billed Authorized Consumption	Billed Metered Consumption	Revenue Water
			Billed Unmetered Consumption	
		Unbilled Authorized Consumption	Unbilled Metered Consumption	Non-Revenue Water
			Unbilled Unmetered Consumption	
	Water Losses	Apparent Losses	Unauthorized Consumption	
			Customer Meter Inaccuracies	
			Data Handling and Billing Errors	
		Real Losses	Leakage on Transmission and/or Distribution Mains	
			Leakage and Overflows at Utility's Storage Tanks	
			Leakage on Service Connections up to Point of Customer Metering	

Figure 2.2. IWA standard water balance (Lambert and Hirner 2000)

Meanwhile, the term "real losses" gives an impression that other types of losses are not real. It could also be referred to as "Leakage," as it is more self-defined. Leakages occur as bursts or background leaks on mains, service connections, and reservoirs. The term "system input volume" could also be referred to as "input water" for straightforward wording with the same considerations that are applied to the calculation of the system input volume. Such revisions help unfamiliar scientists, outsiders, and newcomers to better understand the essence of WL components and engage in its research and discussions. In addition, the term "non-revenue water" should not replace the name of the discipline, "water losses", in water distribution systems. This is because the term NRW is dominated by a commercial sense, whereas "water loss" belongs to the field of water conservation and demand management, with a nobler and wider goal than just increasing the revenues of water utilities. The environmental perspective should remain in the general notion of the topic, even if it is also technical and commercial.

2.3 WATER LOSS ASSESSMENT METHODS

WL assessment can be conducted in several stages, as shown in Figure 2.3. The first stage is to determine how much the total volume of WL is (Figure 2.3a). This can be calculated directly from Equations 2.1 and 2.2:

$$NRW\ (\%) = \frac{SIV\ (m^3/yr) - BW\ (m^3/yr)}{SIV\ (m^3/yr)}$$
$$= AL\ (\%) + RL\ (\%) + UAC\ (\%)$$

(2.1)

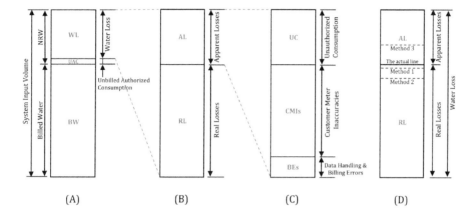

Figure 2.3. Water loss assessment stages

$$WL = NRW - UAC \qquad\qquad (2.2)$$

where NRW is non-revenue water; SIV is system input volume; BW is billed water; WL is water loss; AL is apparent losses; RL is real losses; and UAC is unbilled authorised consumption, which is usually a small component that can be estimated from records of the water utility. The second stage is to breakdown the total volume of WL into its two components: apparent losses and real losses (Figure 2.3b). The third stage is to conduct sub-component analysis typical for apparent losses (Figure 2.3c) and also for real losses. An advanced stage is then conducting an accuracy assessment and/or comparative analysis of the two main components of WL, when more than one WL assessment method can be applied or integrated (Figure 2.3d).

The crucial step in this scheme is the second stage, at which the total volume of WL is broken down into its two main components and the line between apparent losses and real losses is drawn (Figure 2.3b). There are three common methods for the component estimation process: Minimum Night Flow Analysis (MNF), Bursts And Background Estimates (BABE), and Top-Down Water Balance. While MNF is a field-based method, BABE and Water Balance are desk methods. The following sections explain in detail these methods.

2.3.1 Minimum night flow analysis

MNF analysis estimates the real losses in a separated small part of the network. Once real losses are estimated, the apparent losses can then be calculated by subtracting the volume of real losses from the total volume of WL.

An MNF analysis is usually performed in a District Metered Area (DMA), which is a hydraulically isolated part of the network. DMA is a discrete zone with a permanent boundary defined by flow meters and/or closed valves (Farley and Trow 2003). It typically encompasses between 500–3000 customer service connections with a measured supply input flow (AWWA 2009; Thornton et al. 2008). DMAs may either be already established in the distribution system or temporal DMA is to be established to undertake MNF analysis (Fanner 2004). Elaborate considerations on the design and establishment of DMAs are provided in several publications (AWWA 2009; Chisakuta et al. 2011; Di Nardo et al. 2013a; Farley and Trow 2003; Farley et al. 2008; Galdiero et al. 2015; Kesavan and Chandrashekar 1972; Morrison et al. 2007; Thornton et al. 2008).

The MNF is the lowest flow into the DMA over a 24 hour period as shown in Figure 2.4. Through MNF analysis, estimation of real losses is conducted by analysing 24-hour zone measurements to determine the MNF that normally occurs between 02:00 and 04:00 AM, during which most users do not use water or are inactive. Therefore, the water flow during this time of the day is predominantly represented by leaks (Farley and Trow 2003; Liemberger and Farley 2004; Puust et al. 2010).

The estimation of the real loss component through this method is carried out by subtracting the possible legitimate night usage from the MNF through Equation 2.3 (Chisakuta et al. 2011; Farley and Trow 2003):

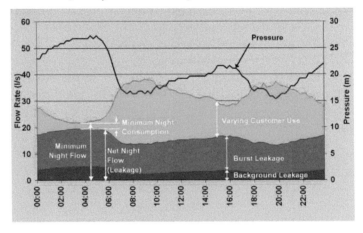

Figure 2.4. Variation of flow (indicating MNF), pressure, and leakage in a DMA (Courtesy of Lambert A, Liemberger R. and Thornton J.).

$$L_{DMA@t_{MNF}} = Q_{MNF} - Q_{LNC} \qquad (2.3)$$

where $L_{DMA@t_{MNF}}$ is the leakage rate in the DMA (m^3/h) at the time of MNF, Q_{MNF} is the minimum flow rate, and Q_{LNC} is the legitimate night consumption in the DMA at the time of MNF. Q_{LNC} should be accurately estimated case by case but can be roughly calculated based on an assumption that 6% of the population are active and the water use is for toilet flush on the order of 1-5 litres per property per hour, depending on the number of persons per household and size of the toilet cistern (Fantozzi and Lambert 2012; Hamilton and McKenzie 2014).

Nonetheless, the result obtained from Equation 2.3 indicates real losses at the time of MNF only and not for the entire day. Estimating the real loss value through generalising the $L_{DMA@t_{MNF}}$ for all hours of the day would lead to an overestimation of the daily leakage, because of the lower average pressure during the day (due to higher flows). The average pressure and, thus, the leaks in the DMA changes over a 24-hour period, depending on the pressure pattern of the supply system (AWWA 2009; Farley et al. 2008; Thornton et al. 2008). When the DMA has its lowest inflows, the pressure is at its highest and, thus, are the leakages, as shown in Figure 2.4. For this reason, the MNF leakage should be modelled according to the leakage–pressure relationship.

In principle, a leak from an orifice in a rigid pipe can be calculated based on the Torricelli equation, presented in Equation 2.4. This equation presents a square-root relationship between the leakage and the head of the water. Equation 2.4 cannot be used for non-rigid pipes that can split in which the area of split varies exponentially with pressure. For this reason, Van Zyl et al. (2017) suggested a modified version of the orifice equation, where the fixed orifice area and flexible orifice area are considered, as shown in Equation 2.5.

$$Q_O = C_d A_0 \sqrt{2gh} \qquad\qquad (2.4)$$

$$Q_{OM} = sgn(h) C_d \sqrt{2g} (A_0 |h^{0.5}| + m|h^{1.5}|) \qquad\qquad (2.5)$$

where Q_O is the orifice flow rate; Q_{OM} is the modified orifice flow rate; C_d is the discharge coefficient; A_0 is the (initial) orifice area; g is the acceleration due to gravity; h is the pressure head differential over the leak opening ($h_{internal} - h_{external}$), in which a leak occurs only if the quantity is positive; sgn is the sign function, allowing the consideration of leakage out of the pipe (+) as well as intrusion into the pipe (−); and m is the head-area slope. Earlier, empirical studies applied this concern in the Fixed and Variable Area Discharges (FAVAD) principle, which demonstrates that most discharges from pressurised pipelines vary with pressure to a greater or lesser extent. The leakage exponent N_1 is accordingly introduced by May (1994) and Lambert (2001). N_1 in Equation 2.6 varies from 0.5 for a fixed area to 1.5 for a flexible area (Lambert 1997; Lambert 2001; May 1994). Later studies proposed a wider range of N_1 values, e.g.,

21

between 0.36 and 2.95 (Farley and Trow 2003; Greyvenstein and van Zyl 2007; Lambert 1997; Schwaller et al. 2015; Ssozi et al. 2016; Van Zyl et al. 2017).

$$Q_i/Q_{MNF} = (P_i/P_{MNF})^{N_1} \qquad (2.6)$$

where Q_i is the leakage rate, P_i is the average pressure in the DMA during time i, and Q_{MNF} and P_{MNF} are the leakage rate and average pressure at the time of MNF, respectively.

Fixed area leakage usually occurs in rigid pipe materials, and variable area leaks occur in flexible pipe materials, such as PVC or polyethylene, which can split and where the area of the split also varies with pressure. Accordingly, the N_1 exponent is 0.5 for rigid pipes and 1.5 for flexible pipes, assuming values in between for a mixed-pipe network (Lambert 2001; Puust et al. 2010). McKenzie et al. (2003) reported that from various tests conducted around the world, the average N_1 value for a system is of the order of 1.15 and could be assumed to be 1.0, implying a linear relationship between leak flow rates and pressure, unless information is available to calculate the true value from recorded data (McKenzie et al. 2003; Morrison et al. 2007). Nevertheless, estimating the relationship between leakage exponent N_1 and the fluctuating pressure in the DMA during the day is increasingly discussed (Di Nardo et al. 2015; Lambert et al. 2017a; Laucelli and Meniconi 2015; Van Zyl and Cassa 2014). The zonal night test is used to determine the variable N_1, which is influenced by a changing pressure in the DMA. This test is only possible when the legitimate night consumption is minimal and the MNF in the DMA is near the leakage rate.

Therefore, with knowledge of the pressure–leakage relationship and the value of the exponent N_1, the volume of the leakage can be calculated at any hour of the day from Equation 2.6. Nevertheless, to get the daily leakage volume (or real losses), a night–day factor (F_{ND}) should be calculated. F_{ND} is a parameter that relates the night leakage rate to the daily leakage rate as shown in Equation 2.7 (Morrison et al. 2007). F_{ND} can be calculated using Equation 2.8 or alternatively Equation 2.9 (Lambert et al. 2017a; Morrison et al. 2007; Pillot et al. 2014).

$$Q_{DMA;\,daily} = L_{DMA@t_{MNF}} \times F_{ND} \qquad (2.7)$$

$$F_{ND} = \sum_{i=0}^{23} (P_i/P_{MNF})^{N_1} \qquad (2.8)$$

$$F_{ND} = 24 \times P_{avg.daily}/P_{MNF} \qquad (2.9)$$

where $Q_{DMA;daily}$ is the daily leakage in the DMA; F_{ND} is the night–day factor (rate per hour); i is the hour of the day, starting from 0 hour to the last hour in the day, the 23rd hour, for a total of 24 hours; $P_{avg.daily}$ is the averagae daily pressure in the DMA; and P_{MNF} is the pressure at the time of MNF.

F_{ND} is usually less than or equal to 24 h/day for DMAs with gravity supply and can reach as low as 12 h/day for low-pressure gravity systems with large frictional head losses. For DMAs supplied by direct pressure, the F_{ND} is usually greater than 24 h/day and can be as high as 36 h/day. In gravity-fed systems, values of 18 to 24 are typical (Morrison et al. 2007).

Meanwhile, Equation 2.9 estimates F_{ND} using P_{avg}. Wu et al. (2011) and Lambert and Taylor (2010) have presented a systematic approach to calculating the average pressure at any supply system through the following: (i) calculating the weighted average ground level of the zone; (ii) near the centre of the zone, identify a convenient management point that has the same weighted average ground level. This point is known as the average zone point (AZP); and (iii) measure the pressure at the AZP and use this as the surrogate average pressure for the zone. AZP pressures should be calculated as average 24 hour values, as the average pressure is a sensitive parameter in estimating the real losses through MNF analysis (Høgh 2014).

Knowing the value of F_{ND} from Equation 2.8 or Equation 2.9, the total volume of real losses in a specific DMA can be estimated through Equation 2.7. Nonetheless, accuracy of MNF analysis depends on several technical and estimation issues. Application and accuracy considerations of MNF analysis is elaborated in the Supplementary Information of this chapter, Werner et al. (2011), Mimi et al. (2004), Alkasseh et al. (2013), Salim and Manurung (2012), Fantozzi and Lambert (2010), and Hamilton and McKenzie (2014).

2.3.2 Burst and background estimates

The Underlying Concept

The BABE approach was first introduced by Lambert (1994). It was the first approach to model leakage components objectively, rather than empirically (AWWA 2009). Through the BABE approach, WL is assessed by estimating the volume of real losses; then, the apparent losses can be calculated by subtracting the volume of real losses from the total volume of WL.

The underlying principle of BABE concept is that real losses consist of numerous leakage events. The loss volume for each event is a function of the average flow rates and average runtimes for different types of leakages (Thornton et al. 2008). In the BABE concept, the volume of an individual leak or burst is calculated as the average flow rate times the duration during which the leak or burst runs, as shown in Equation 2.10.

$$V = Q \times T \qquad\qquad (2.10)$$

where V is the volume of the leakage; Q is the leak flow rate; and T is the leak duration. Based on Equation 2.10, it can be deduced that Lambert and Morrison (1996) categorised leakages into categories related to the two parameters on the right-hand side of Equation 2.10—flow-rate-based categories and duration-based categories.

In keeping with Lambert (1994), the range of flow rates from leaks to bursts is immense. Based on the flow rate criteria, a leak is either a burst with a high flow rate and, thus, should be reported or detected by the utility or a background leak with a low flow rate that cannot be detected by the utility. Therefore, individual points of loss have been categorised as background losses, unless the flow rate is at least 500 L/h, in which it is categorised as bursts. Therefore, almost all losses from fittings in mains and service connections (including air valves, hydrants, stop taps, dripping taps, cisterns, etc.) fall within the background category. For tanks and reservoirs, background losses represent leakage from the structure, and overflows are equivalent of bursts.

Meanwhile, the duration of a leak depends on the policies of the water utility, to what extent the utility conducts leakage detection campaigns to repair detectable leaks, and how quick the response of the utility is to repair a leak once the utility is aware of the leak either when reported by public or detected by its manpower campaigns. In this sense, background leaks are always continuous. Hence, while the duration of reported bursts is related to the standards of service and repair policies, the duration of unreported bursts is related to the method of active leakage control (ALC) practiced by the utility.

Later definitions for types and duration of leakage are as follows: (a) background losses are the aggregation of small leaks with flow rates too low (≤ 0.5 m^3/h) to be detected by an ALC or leak detection survey of the utility; (b) reported bursts are visible and usually quickly reported by the public or observed by the water utility staff; and (c) unreported bursts are leaks that are not visible at the surface but are usually discovered during leak detection surveys (AWWA 2009; Farley et al. 2008; Lambert and Morrison 1996).

While background losses are expected to run continuously, reported and unreported bursts have variable durations (Lambert and Morrison 1996). As shown in Figure 2.5, the burst duration can be divided into three components: (i) Awareness Time: the time from the occurrence of a leak until the water utility becomes aware of its existence; (ii) Location Time: the time it takes a water utility to investigate the report of a leak and correctly locate its position so that a repair can be performed; and (iii) Repair Time: the time it takes a water utility to repair a leak once a leak has been located (AWWA 2009; Farley et al. 2008; Wu et al. 2011).

Accordingly, the BABE approach uses data from three distinct sources: standard components (e.g., pressure correction and average burst flow rates); auditable local data (e.g., length of mains and frequency of bursts); and data from company policies in terms of their influence on the duration that a burst runs (Lambert 1994).

Utilising the above classifications, the BABE approach involved the use of Equation 2.11 to model several parameters but for different groups of reported and unreported bursts.

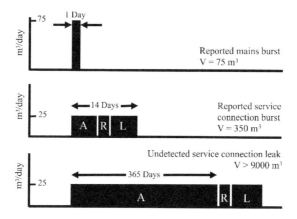

Figure 2.5. Leak run time and volume of water loss; A: Awareness Time; L: Locating Time; R: Repair Time. Adapted from Lambert (1994) and Farley et al. (2008).

$$V = N \times Q \times T \qquad\qquad (2.11)$$

where V is the volume of leakage; N is the number of leaks; Q is the leak flow rate; and T is the average leak duration. To generate factors from the BABE model for practical use, actual data were used to empirically model leakages using a particular case study, in which it is assumed that all bursts $\geq$ 500 l/h have been temporarily shut off or repaired. The model has many assumptions that are elaborated in the Supplementary Information of this chapter. The model assumptions that were originally introduced in Lambert (1994) were fine-tuned in Lambert et al. (1999), Lambert and McKenzie (2002), and Lambert (2009). Ultimately, typical factors were generated to simplify the calculation of the total volume of leakage.

Estimating Real Losses Using BABE Factors

To estimate real losses using the BABE factors, the avoidable and unavoidable annual real losses should be estimated and aggregated.

Table 2.1 presents typical flow rates used to estimate the avoidable volume of leakage from reported and unreported bursts using data from the utility concerning the number of bursts from mains and service connections that are reported by public and the number of bursts from mains and service connections that are detected through leak detection surveys conducted based on the utility ALC policy (AWWA 2009; Farley et al. 2008; Wu et al. 2011).

Table 2.1. Flow rates for avoidable reported and unreported bursts at 50 m pressure (Farley et al. 2008).

Location of Burst	Flow Rate for Reported Bursts [l/h/m pressure]	Flow Rate for Unreported Bursts [l/h/m pressure]
Mains	240	120
Service Connection	32	32

Table 2.2 shows factors used to estimate the unavoidable background, reported, and unreported losses using the number of service connections and length of mains. The result should then be adjusted to the value of the average pressure of the entire network, such that the total volume of unavoidable background losses can be estimated (AWWA 2009; Farley et al. 2008; Wu et al. 2011). It is worth mentioning that this part of the model represents the unavoidable annual real losses (UARL) that can be estimated separately from Equation 2.12 (Lambert et al. 1999; Lambert et al. 2014), which is used to calculate the infrastructure leakage index, a widely accepted leakage performance indicator.

$$UARL = (18 \times \frac{L_m}{N_c} + 0.80 + 0.025 \times L_P) \times P_{avg} \qquad (2.12)$$

where UARL is the unavoidable annual real losses (L/service connection/day); L_m is the length of mains in km; N_c is the number of service connections; L_p is the total length in km of underground connection private pipes (between the edge of the street and customer meters), and P_{avg} is the average operating pressure in metres. Recent feedback from Allan Lambert indicates that Equation 2.12 should be considered only for unavoidable background losses, which usually account for 67% of the UARL. Other unavoidable reported and unreported bursts can be estimated using Equation 2.9 and then added to results of Equation 2.12.

The total volume of real losses can eventually be estimated by aggregating the avoidable and unavoidable losses. Note that the factors generated by the BABE model are easy to use. However, the assumptions of the model, in the Supplementary Information of this chapter, should be checked before assessing WL through this method. In addition, there is a calibration factor for the unavoidable background leakage, which considers the conditions of the infrastructure of the system to be assessed compared with the conditions of the infrastructure of the cases for which the model factors have been developed. This factor is called the "infrastructure condition factor," and its value varies from 1 to 3. More information on this factor can be found in Fanner and Thornton (2005).

Table 2.2. Components of unavoidable annual real losses at a pressure of 50 m (Lambert 2009).

Infrastructure Component	Unavoidable Background Leakage (UBL)	Reported Breaks	Unreported Breaks	Unavoidable Annual Real Losses (UARL)	
Mains	480 litres/km/day	290 litres/km/day	130 litres/km/day	900 litres/km/day	18 litres/km/day/ metre of pressure
Service Connections (mains to curb-stop)	30 litres/conn/day	2 litres/conn/day	8 litres/conn/day	40 litres/conn/day	0.8 litres/conn/day/ metre of pressure
Service Connections (curb-stop to meter)	800 litres/km/day	95 litres/km/day	355 litres/km/day	1250 litres/km/day	25 litres/km/day/ metre of pressure
Typical FAVAD N_1	Close to 1.5	0.5 to 2.5 depends on pipe materials and types of leaks		Assumed as average of 1.0 for UARL formula	

2.3.3 Top-down water balance

The Top-Down Water Balance was first introduced by Lambert and Hirner (2000) in the UK and internationally by Lambert (2002). Unlike the MNF analysis and BABE approach, the components of the apparent losses are estimated first in the water balance methodology; then, the real losses can be calculated by subtracting the volume of the apparent losses from the total volume of WL.

According to the IWA standard water balance, all water should be quantified, via measurement or estimate, as either authorised consumption or losses. The top-down approach to conduct a water balance contains four basic steps, as demonstrated in AWWA (2009), Farley et al. (2008), Farley and Trow (2003), Alegre et al. (2000), and Lambert and Hirner (2000):

(1) Determining the system input volume: the amount of produced and/or imported water.

(2) Determining the authorised consumption: (i) billed: total volume of water billed and sold by the water utility and (ii) unbilled: total volume of water provided at no charge; (metered and not metered)

(3) Estimating the apparent losses: (i) theft of water and fraud; (ii) meter under-registration, since the customer meters tend to be under-registered rather than over-registered (AWWA 2009); and (iii) data handling errors

(4) Calculating the real losses, then trying to classify them as follows: (i) leakage on transmission mains; (ii) leakage on distribution mains; (iii) leakage from reservoirs and overflows; and (iv) leakage on customer service connections

According to these steps, system input volume, billed consumption, and unbilled metered authorised consumption are usually metered. In contrast, the unbilled authorised unmetered and apparent losses are estimated. The unbilled authorised consumption (metered and unmetered) is usually a small component and, thus, typically assumes a range from 0.5% (Lambert and Taylor 2010) to 1.25% (AWWA 2009) of the system input volume or is estimated by the utility, as it is case-specific.

Meanwhile, the apparent loss estimation starts with assuming the unauthorised consumption at 0.25%, as in AWWA (2009), or 1%, as in Lambert and Taylor (2010), or it could be assumed at 10% of the billed water for developing countries, as suggested by Mutikanga et al. (2011a). Alternatively, it could also be estimated via the experience of the utility with validated data (AWWA 2009). Afterwards, the customer meter inaccuracies should be estimated according to meter tests at different flow rates representing typical customer water use and meter guidance manuals (Arregui et al. 2006a; AWWA 2009; Farley et al. 2008; Mutikanga et al. 2011b). The next step is to estimate the systematic data handling errors by exporting and analysing historic billing data trends for a certain period (Farley et al. 2008; Mutikanga et al. 2011a). Eventually, the component of apparent losses is estimated by summing its subcomponents. The real losses are then calculated by subtracting the apparent losses from the total volume of WL. Following these steps, the WL components are quantified, and the water balance is established through the top-down approach. The results of this methodology are usually presented in the standard form in Figure 2.2. For satisfactory results, Alegre et al. (2000) recommended that all water balance calculations should be associated with confidence grades to improve the reliability of the sensitive parameters. To improve the reliability of the water balance estimates, components with the greatest variance should be the priority (Lambert 2003).

2.4 OTHER WATER LOSS ASSESSMENT METHODS

In the case of the availability of regular and abundant data as well as the sufficient technical capacity of a water utility, the hydraulic model of the network can be used to estimate, quantify and detect real losses. Examples of such models can be found in Palau et al. (2012), Tabesh et al. (2009), Giustolisi et al. (2008), Almandoz et al. (2005), and Buchberger and Nadimpalli (2004). Management models, including recent tools and models, are presented in Mutikanga (2012). Recent top-down approaches and models exist. The evaluation of WL components through conducting water and wastewater (mass) balance without the need for pressure-dependent models or methods is suggested in Al-Washali (2011). This approach utilises the fact that apparent losses reach the wastewater treatment plant, whereas real losses do not. Al-Omari (2013) applied the concept and equations developed in Al-Washali (2011) in a water evaluation and planning model and evaluated the components of WL in the cities of Amman and Zarqa in Jordan.

2.5 SUB-COMPONENT ANALYSIS

After the components of apparent losses and real losses are estimated, and the line between apparent losses and real losses in Figure 2.3-b is drawn, further sub-component analysis should be carried out. While sub-components of real losses can be assessed only through BABE approach or factors, there are several methods for breaking down the apparent losses into sub-components. Sub-component assessment of apparent losses is conducted though evaluating customer meter inaccuracies, data handling and billing errors and unauthorized consumption (AWWA 2009; Farley et al. 2008; Thornton et al. 2008; Vermersch et al. 2016). Due to the fact that apparent losses are not a dominant component in developed countries, it is common that they are assumed at standard percentages as in AWWA (2009) and Lambert and Taylor (2010). For European utilities, Lambert et al. (2014) recommended default values of apparent losses. However, such assumptions are not applicable for developing countries that have different context. Several methods of assessment of customer meter inaccuracies and other components of apparent losses are presented in Arregui et al. (2006a), Arregui et al. (2015), Claudio et al. (2015), Ncube and Taigbenu (2019), Mutikanga (2012), AWWA (2009), Criminisi et al. (2009), Farley et al. (2008) and Seago et al. (2004).

2.6 DISCUSSION AND CONCLUSIONS

Several WL assessment methods are reviewed in this chapter. The crucial step in WL assessment is breaking down the total volume of WL into its two components—apparent losses and real losses (Figure 2.3b). There are three common methods for component estimation: MNF, BABE, and the Top-Down Water Balance.

- MNF remains the only method that provides valuable actual measurements whose accuracy can be evaluated. It helps a utility to downscale WL assessment to a level that enables a utility to better manage its network and control leakage. However, MNF requires intensive fieldwork that limits its use for regular assessment or baseline assessment. MNF needs investment, sophisticated equipment, and advanced technical awareness of the network components. The accuracy of this method depends on the technical capacity of the utility manpower, the average pressure to which the calculations are sensitive, and an estimation of the legitimate night consumption, which, in turn, is a sensitive parameter that is highly influenced by the population density and consumption habits. To generate the output of this method and generalise it for the entire network annually, it should be conducted regularly throughout the year and for several representative DMAs.

- Assessing real losses using the factors generated by the BABE model is a straightforward step that uses the available data from a utility in a developed country. The BABE concept is the only approach that breaks down real losses into sub-components, allowing the utility to better understand the nature and types of leaks in the network and realise the impacts of the utility-leak-related policies on the magnitude of real losses. The concept of the model also allows the consideration of case-specific conditions of the infrastructure and evaluate the management practices toward real losses. However, there should be a differentiation between the BABE concept that includes the UARL and the BABE factors that are used to assess WL. Assessing real losses through BABE factors is questionable. The model uses many assumptions from specific cases that may not be sufficiently representative for various international networks. There is neither a complete presentation of the development of the model equations and factors nor the availability of complete data for the calibration and validation of the model for other cases in developed or developing countries. The model should not be applied to utilities with no regular ALC, such as those in developing countries. WL assessment should not be conducted through BABE factors unless there is no other option, owing to its many assumptions that lead to the underestimation of the volume of real losses. Instead, this method should be used as a supplementary tool to breakdown the volume of real losses into its sub-components. The BABE concept including the UARL principle should be promoted for stages beyond WL assessment, such as WL reduction and management.

- The Top-Down Water Balance is neither pressure-dependent nor an extensive-fieldwork method. It is a cost-effective assessment that should be used first and conducted annually, allowing regular internal and external monitoring of real losses. However, its assumptions related to apparent losses are not always applicable to the various water utilities, particularly those in developing countries. The lack of an objective methodology for estimating unauthorised consumption is a major limitation. The principle of assuming specific sub-components of WL could be negatively influential. Whenever a sub-component is always assumed at a certain level, it cannot be monitored any more toward its reduction measures. This would be a critical issue for utilities for which apparent losses are significant and regular monitoring of the level of apparent losses is a priority.

- Table 2.3 summarises the advantages, limitations, and potential use of each WL assessment method. While MNF analysis is appropriate only for systems with zoned networks and where several DMAs are established, the water balance should be the first option to be used in general. In contrast, BABE factors should not be used on its own.

- There are some research gaps that should be tackled. There is a need for guidance notes on accuracy and application considerations for MNF analysis, namely the identification of pressure-leakage relationship, estimation of legitimate consumption and practical technical and technological considerations. The BABE model needs calibration and validation in more international, representative, and typical cases. The model also needs an adaptation study of its assumptions and factors to a developing country context. Revising the infrastructure condition factor may help bridge the gap between the output of this method and the actual situation in many cases in developing countries. The Top-Down Water Balance is a promising method. However, it requires several improvements. There is a demand for more objective assumptions of apparent losses, particularly for developing countries, and development of practical methods to estimate the unauthorised consumption. Improved estimation of unauthorised consumption assists in the formulation of more accurate water balance as well as planning of effective water loss management in distribution networks.

Table 2.3. Pros and cons of water loss assessment methods.

Method	Advantages	Limitations	Requirements	Accuracy Issues	Potential Use
MNF Analysis	Actual measurements Both assessment and reduction process	Intensive field work Requires trained manpower Pressure-dependent	Sophisticated equipment Network zoning Technical capacity	Logged data are prone to technical mis-installation Representativeness of the DMA	For cases where DMAs are established, GIS or hydraulic models exist, and where a DMA could be representative for the whole network
BABE Factors	Straightforward The only method that breaks down real losses into sub-components	Pressure-dependent Intensive assumptions	Pressure measurements Maintenance records	Unrepresentative assumptions Average pressure of the whole network is always questionable and rarely accurate	Applicable for utilities with ALC Should not be applied unless there is no other option.
	Considers the infrastructure conditions	The model factors are not calibrated for other case studies in developed or developing countries	Date of lengths of mains and Nr. of customer connections	The method likely underestimates the volume of real losses	Should be applied as a supplementary method to break down real losses
	Clarifies the nature of leakages	Applicable only for utilities that have regular Active Leakage Control		Low accuracy in general	Its principle of UARL can be used to consider the infrastructure conditions
Water Balance	Non-pressure-dependent Cost-effective	Focuses on real losses more than apparent losses The assumption approach freezes the assumed variable and so cannot be monitored any more	Available data Estimation of unauthorized consumption	Method accuracy depends on the accuracy of the apparent losses assumptions. This should be critical	Should be used for developed countries. For developing countries, it can be used once more accurate assumptions exist or unauthorized consumption can be estimated
	Clear and improvable assumptions	Inappropriate assumptions of apparent losses for developing countries No methodology for estimation of unauthorized consumption	Estimation of meter inaccuracies	The method likely overestimates the real losses	

2.7 SUPPLEMENTARY INFORMATION

2.7.1 Application and interpretation of MNF analysis

The application of MNF analysis undertakes several considerations. Werner et al. (2011) presented their experience of installing 200 Magflow meters to assist 75 local water utilities to conduct night flows measurements, monitoring flows into DMAs. To obtain satisfactory results from the MNF tests, they suggested some recommendations including the following: (i) meters must be sized correctly to achieve accuracy at low flows but without creating excessive head loss. The meter often needs to be smaller than the pipe feeding the area. Otherwise, obtained data would indicate real losses that were too low, as shown in Figure 2.6a; (ii) the use of in-reservoir metering can be cost-effective if the alternative is multiple meters on the lines outside the reservoir; (iii) logging flows should have a frequency of 15 min, with the average flow calculated from the pulse outputs. The data should be converted to hourly averages; and (iv) other technical problems would affect the entire accuracy of the analysis. Figure 2.6b shows the outputs obtained from two reservoir outlet meters feeding into a common zone where the meters had not been calibrated since installation and unshielded wire had been used to connect the sensor to the transmitter, resulting in interference and inaccurate readings. Further practical considerations are available in Werner et al. (2011). These recommendations indicate how the accuracy of MNF analysis can be significantly influenced by many application factors.

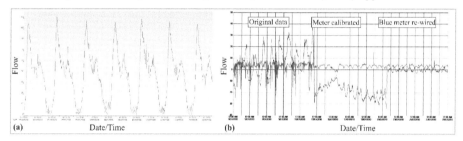

Figure 2.6. Field flow measurements: (a) over-sized meter yielding low-accuracy measurements at MNF, indicating low real losses; (b) logged outputs of two meters that were out of calibration and inaccurate (Werner et al. 2011).

Mimi et al. (2004) emphasised that MNF should be recorded for not less than one week to ensure that the reading will repeat itself as it varies during the days of the week. Analysing the data of 30 DMAs, Alkasseh et al. (2013) found that MNF occurs between 1:00–5:00 AM in Malaysia. Salim and Manurung (2012) found that application of MNF under real conditions can be relatively complex and results can be misleading, particularly in areas where night consumption is high especially in dense population areas. For better accuracy of MNF analysis, Fantozzi and Lambert (2012) showed that measurements of night consumption do not follow a statistical normal distribution and recommended the use of percentage of active population during MNF with a binomial distribution. This implies how the estimation of night consumption influences the accuracy of the analysis.

Hamilton and McKenzie (2014) presented DMA data of a typical zone that experienced intermittent supply with only one complete day of supply. As shown in Figure 2.7a, when the zone is subjected to regular periods of pressure followed by periods of no pressure, the leakage levels in the zone tend to become very large. The one day during which the water was not cut off provides valuable information on the minimum night flow and, thus, the level of leakage in the system.

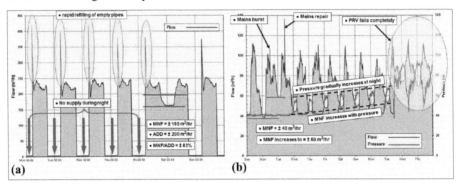

Figure 2.7. Field flow measurements: (a) 7 day MNF data for intermittent supply into a DMA; (b) MNF data at pressure-reducing valve (Hamilton and McKenzie 2014).

An example on how to interpret MNF data is provided by Hamilton and McKenzie (2014) as shown in Figure 2.7b. The graph is of data from the city of Johannesburg for a pressure reducing valve (PRV) installation. The interpretation of such a figure should start at the left going towards the right, and the following could be concluded from the graph: (i) the zone initially experiences a relatively high night flow of 40 m^3/h; (ii) night flow jumps by 20 m^3/h on day 2 because of a mains burst, which pushes the minimum night flow up to 60 m^3/h; (iii) the leak is repaired during day 3, as indicated by a refilling spike and the drop in the minimum night flow back to 40 m^3/h; (iv) over the next six nights, the PRV starts to experience problems and is unable to maintain a fixed pressure; (v) during the day, the pressure appears stable due to the higher demand; (vi) the minimum night flow also gradually increases each night in response to the higher pressures; and (vii) after the six nights of gradually higher pressures due to the failure of the PRV, it fails completely and no longer provides any pressure control. The minimum night flow rises significantly in response to the higher pressure. This is an example of how MNF graphs should be analysed and how pressure influences the level of leakage.

2.7.2 Assumptions of BABE model

Despite later fine-tuning to some of the BABE assumptions, as in Lambert et al. (1999) and Lambert and McKenzie (2002), it can be concluded from Lambert (1994) that the following assumptions are associated with the factors generated by the BABE model:

Flow rates

- Flow rate is assumed as 25 m^3/day for an underground service connection burst at 40 m pressure.
- Flow rate is 75 m^3/day for typical distribution mains burst at 40 m pressure.
- Flow rate is 150 m^3/day for typical trunk mains burst at 40 m pressure.
- These flow rates increase and decrease according to the specific pressure–leakage relationship.

Reported bursts

- All reported bursts will be repaired within specific durations.
- A reported trunk mains burst will be repaired in 1 day (Awareness (A) + Location (L) + Repair (R) = 1).
- A reported distribution mains burst will be repaired in 1.1 days (A + L + R = 1.1).
- A service connection burst will be repaired in 16 days (A = 4, L = 2, and R = 10).
- A private pipe burst will be repaired in 46 days (A = 4, L = 2, and R = 40). The reason for segregating the private pipe from the service connection is that the leak detection surveys do not investigate areas beyond the boundary of private ownership (e.g., inside backyard or open areas of a customer house). Therefore, a burst in the private pipe is either repaired by the customer or often requires long notice procedures.

Unreported bursts

- All unreported bursts will be repaired within specific durations.
- Time required for unreported burst on mains to be detected through ALC and, thus, located right away and repaired later is 195 days (A = 183, L = 0, R = 12).
- Time required for unreported bursts on service connection to be detected through ALC and, thus, instantly located and repaired is 267 days (A = 253, L = 0, R = 14).
- Time required for unreported bursts on private pipe to be detected and repaired is 297 days.

Other assumptions

- ALC is conducted by the utility, and regular leak detection survey occurs once a year. The detection technology and technical capacity of manpower are as those in the case in which the model is developed.
- Length of mains is used as a surrogate for the numbers of fittings on the mains.
- Number of service connections is used as a surrogate for the number of fittings on the service connections.

- A burst is a leak with flow rate $\geq$ 500 L/h. Otherwise, it is considered background. Background losses for the service connection are split 67% to 33% between the service connection and the private pipe
- Fittings losses from the private pipe are considered to be consumed by the customer
- As the trunk mains and reservoirs are not included in the input data of the BABE model, it is assumed that background losses for trunk mains are 0.2 m^3/km trunk mains/y for each year of age. For reservoirs, it was assumed that the background losses are 33% of capacity per day.
- All parameters of the BABE model are insensitive except for the pressure and duration of the bursts.

3

OVERVIEW OF NON-REVENUE WATER ASSESSMENT SOFTWARE TOOLS

Several software tools are available that can assess the performance of non-revenue water (NRW) in water distribution networks and plan for reduction measures. Of the 21 tools that have been reported in the literature, 12 are freely available. The creation of these many tools and different versions of each individual tool indicates the promising future of NRW software development. This overview comprises 12 freely available tools for water balance establishment, NRW performance assessment and NRW reduction planning. Most of the tools have been developed to establish standard annual water balances and recommend performance indicators for the entire network. Some tools have been developed to intervene and reduce the leakage in a district metered area (DMA). Key features increasingly being included in NRW software include uncertainty analysis, recognition of supply intermittency, and accommodation of a guidance matrix and benchmarks. Leakage assessment is fully recognised, and leakage reduction analyses are increasingly growing in the software tools. However, much less attention has been paid to assessing and options for reducing apparent losses. Although a comprehensive NRW management tool for monitoring, planning and intervention is not currently available, developing a comprehensive tool is worthwhile, in the form of one package or a kit of smaller tools. Towards this goal, this chapter provides insights and recommendations addressing topics of intermittency, normalisation, multi-method assessment, planning for the reduction of apparent and real losses, and estimation of the economic level of water loss.

This chapter has been published as: Al-Washali, T., Elkhider, M., Sharma, S., and Kennedy, M. "A Review of Non-Revenue Water Assessment Software Tools." WIREs Water, 7:e1413(02), 2020.

3.1 INTRODUCTION

NRW assessment basically involves quantifying losses in a particular system without considering where the losses are actually taking place (Puust et al. 2010). Firstly, the volume and components of NRW should be estimated. For an intermittent supply, NRW should be normalised (Chapter 4). Normalisation is a straightforward task when it involves adjusting the volume of NRW as the supply in the system is continuous (24/7) (AL-Washali et al. 2019a). Secondly, NRW should be broken down into components using different methods (AL-Washali et al. 2016). Following that, NRW components and subcomponents can be prioritised for intervention measures and to minimise losses in the system (Al-Washali et al. 2020b; AL-Washali et al. 2019b). Dividing the network into DMAs is the key to leakage control (Farley and Trow 2003; Galdiero et al. 2015; Kesavan and Chandrashekar 1972). Incorporating this step with pressure management is fruitful, mainly through the proper usage and location of pressure reducing valves (PRVs) (Alonso et al. 2000; Araujo et al. 2006; Vicente et al. 2016). Pro-active leakage control assists water utilities unearth the hidden leaks, using regular leakage detection surveys (Li et al. 2014; Puust et al. 2010; Wu and Liu 2017). The frequency of leakage detection surveys can be identified economically, based on recognition of the rate of rise of leakage (Lambert and Fantozzi 2005; Lambert and Lalonde 2005). The economic level of leakage (ELL) can be reached where the cost to further reduce leakage exceeds the expected benefits (Ashton and Hope 2001; Kanakoudis et al. 2012; Pearson and Trow 2005). A similar concept applies for the economic level of AL (Arregui et al. 2018a). The result of combining the economic levels of AL and RL is the economic level of water loss. Besides the many software tools that simulate and hydraulically model the network pipes and appurtenances such as EPANET, WaterGYMS, InfoWater, WDNetXL, H2O MAPWater, KYPIPE, there are many (commercial) tools particularly designed to assist water utilities assess their losses and plan reduction interventions (Halfawy and Hunaidi 2008; Hamilton and McKenzie 2014; Klingel and Knobloch 2015; Liemberger and McKenzie 2003; Sturm et al. 2014; Tabesh et al. 2009; Tsitsifli and Kanakoudis 2010). This chapter, however, reviews 12 freely available software tools for water loss assessment, investigating their functionalities and limitations, and suggesting guidelines for their use and improvement. This will help software users familiarise themselves with these tools and their underlying concepts, and select the appropriate fit-for-purpose tool for each context. The future prospects for the industry are eventually highlighted.

3.2 NON-REVENUE WATER ASSESSMENT SOFTWARE TOOLS

While this chapter focuses on freely available NRW tools, some tools reported in the literature are commercially available. These include Aquadas QS (Aquadas-QS 2007), Aqualibre (Liemberger and McKenzie 2003), Auditsolve (Sturm et al. 2014), SigmaLite

(ITA 2000), Leaks suite (Lambert 2015b), Prototype (Halfawy and Hunaidi 2008), NAIS (Heydenreich and Kreft 2004), Netbase (Netbase 2019), and ÖVGW spreadsheet (ÖVGW 2009). The 12 freely available tools have been designed for water loss assessment, water balance establishment, and NRW PI evaluation. Table 3.1 presents these tools and their approaches, which were developed in response to the establishment of standard terminology, standard water balance (WB) methods (Lambert and Hirner 2000), and recommended NRW PIs (Alegre et al. 2000). The main functionalities of the tools are: (i) the use of a top-down water audit to estimate or assume AL, from which RL and NRW PIs are calculated. The top-down water balance is usually conducted for a period of one year and encompasses the whole system (i.e. global); (ii) the assessment of the RL based on the bottom-up approach using Minimum Night Flow (MNF) analysis in a DMA; and (iii) the use of the Burst And Background Estimates (BABE) for the whole network and in a DMA-scale. Only one tool- Component Analysis- analyses different leakage reduction options. Table 3.1 summarises these tools and their finicalities. Brief descriptions of each of these tools are also provided in a Supplementary Information of this chapter.

Table 3.1. Free software tools for non-revenue water assessment

Tool (version)	Reference	Developer	Environment	Description	Approach	Scale
AquaLite (v4.5)	(Mckenzie 2007)	WRC	Windows-based	A tool to establish WB and PIs.	Top-down	Global
AWWA Water Audit (v5)	(Water Loss Control Committee 2014)	AWWA	Excel-based	A tool to establish WB and PIs. Uses validity score (qualitative), not uncertainties.	Top-down	Global
BenchLeak	(Mckenzie et al. 2002)	WRC	Excel-based	A tool to establish WB and PIs.	Top-down	Global
BenchLoss (v2a)	(GWR-Ltd 2008)	GWR	Excel-based	A tool to establish WB and PIs.	Top-down	Global
CalacuLEAKator (v4.3)	(Koldžo and Vuc'ijak 2013)	Djevad Koldzo	Excel-based	A tool to establish WB and PIs, based on MNF analyses.	MNF	DMA, Global
CheckCalcs (v6b)	(Lambert 2015a)	ILMSS Ltd	Excel-based	A tool to establish WB and PIs. Provides insights on leakage relationships, N1, N2, N3.	Top-down	Global
Component Analysis	(Sturm et al. 2014)	WRF	Excel-based	WB and PIs are inputs. Analyses potential of leakage reduction interventions.	Top-do., BABE, PM, ALC,..	Global
EconoLeak (v1a)	(Mckenzie and Lambert 2002)	WRC	Excel-based	A tool to establish the ELL with cost-benefit analysis of ALC.	ELL	Global
PresMac (v4.4)	(Mckenzie and Langenhoven 2001)	WRC	Windows-based	Operational tool for pressure management in a DMA, using PRVs.	BABE	DMA
SanFlow (v4.6)	(Mckenzie 1999)	WRC	Windows-based	A tool to model MNF in a DMA and breakdown leakage into components	MNF, BABE	DMA
WB-EasyCalc (v5.16)	(Liemberger and Partners 2018)	Roland Leimberger	Excel-based	A tool to establish WB and PIs. Analyses impacts of changes in pressure, SIV and supply time.	Top-down	Global
WB-PI Calc-UTH (v2.2)	(Tsitsifli and Kanakoudis 2010)	Tsitsifli & Kanakoudis	Excel-based	A tool to establish WB and PIs. Considers the over-billing practices in the balance.	Top-down	Global

3.3 NON-REVENUE WATER ASSESSMENT

Lambert and Hirner (2000) suggested the standard terminology for a standard water balance in water distribution networks, as elaborated in Chapter 2. Deducting the volume of billed consumption (BC) from the system input volume (SIV) gives the volume of the NRW. Deducting the volume of the UAC from NRW gives the volume of water loss. Breaking down water loss into AL and RL involves four methods. The top-down methods start by estimating the volume of AL and then calculating the volume of RL. The bottom-up methods analyse the leakage volume based on field measurements or available records.

In the top-down water audit, the AL components are estimated. To determine customer meter inaccuracies, a representative sample is tested in the laboratory at different flow rates that represent the field conditions (Arregui et al. 2007; Arregui et al. 2006b; Walter et al. 2018). Data handling and billing errors are estimated by investigating historical billing records and trends (AWWA 2016; Mutikanga et al. 2011a; Vermersch et al. 2016). Estimating the amount of unauthorised use is challenging, and therefore it is commonly assumed arbitrarily (Al-Washali et al. 2020b; AWWA 2016; Klingel and Knobloch 2015; Mutikanga et al. 2011a; Seago et al. 2004; Vermersch et al. 2016). After estimating the components of AL, RL can be calculated. Afterwards, the International Water Association (IWA) standard water balance in Figure 2.2 can be established. Another top-down method is the water and wastewater balance method (AL-Washali et al. 2018), which assumes that AL enters the sewer network. Analysing the WWTP inflows and comparing it to the BC enables the estimation of the volume of AL, from which RL are then calculated (Al-Washali et al. 2020b; AL-Washali et al. 2018). These calculations establish the IWA water balance, after which best-practice NRW performance indicators (PIs) can be calculated for target monitoring and leakage benchmarking (Alegre et al. 2016; Alegre et al. 2000). Table 3.2 shows the recommended key PIs of NRW. Historically, the fundamental indicator for monitoring and benchmarking NRW was presenting NRW as a percentage of the SIV, using Equation 2.1. Nevertheless, consistent feedback from field data revealed that relying on this indicator for monitoring and benchmarking NRW progress is rather misleading. This is because it is strongly influenced by water consumption (Lambert et al. 2014), favours less water supply over more supplied water (AL-Washali et al. 2019a), a zero-sum indicator for BC, NRW, and SIV (Lambert 2019), and because when used, the denominator in the first part of Equation 2.1 should be a cause of change in the numerator (Alegre et al. 2016). Tackling this problem, the PIs in Table 3.2 were proposed, to give meaningful input and inform perception about the status of NRW progress. The units of the PIs in Table 3.2 indicate the intuition of each indicator and how it should be calculated.

On the other hand, the bottom-up methods are only for estimating the RL. MNF analysis is carried out for a DMA during night time when most customers are inactive. Flow and pressure measurements are analysed and night flows should indicate the volume of the

RL (Farley and Trow 2003; Puust et al. 2010). Notably, this method can only be carried out for a DMA and scaling it up for the entire network is very uncertain (Al-Washali et al. 2020b; AL-Washali et al. 2019b).

After the RL are estimated through one of the above methods, it can be broken down into its sub-components, using the BABE analysis (Lambert 1994). Although most RL are avoidable, some are unavoidable, even in a new and well-constructed network. Background leaks may be too small to detect by the available detection technology. In contrast, bursts are big enough to be reported for repair by customers or by the utility crew. Unreported bursts are usually detected by the leakage detection surveys (i.e. the Active Leakage Control; ALC) (AWWA 2016; Puust et al. 2010). Unavoidable annual RL (UARL) can be estimated using a recommended empirical equation presented in Equation 2.10 (Lambert et al. 1999; Lambert et al. 2014). The BABE analysis is useful because it enables water utilities to understand the nature of RL and plan reduction measures.

Finally, because the water balance is associated with uncertainties, it is usually accompanied by uncertainty analysis (Lambert et al. 2014; Thornton et al. 2008). The uncertainties of the water balance can be calculated straightforward using the error propagation theory (Taylor 1997). As the water balance problem is a process of adding and subtracting, the general equation of the error propagation theory in Equation 3.1 can be simplified as in Equation 3.2 for addition and subtraction or Equation 3.3 for multiplication and division.

The error propagation analysis is simple and sufficient for the water balance problem. It produces the same results with other advanced methods (Al-Washali et al. 2020b) such as Monte Carlo simulation (Rubinstein and Kroese 2016). However, some of the tools do use the variance analysis based on the statistical principles of the root-mean-square method for the normally distributed data (Thornton et al. 2008), which generates the same uncertainties. In this case, the higher the variance of the water balance component, the more significant its uncertainty becomes. The variance for each water balance component can be calculated using Equation 3.4 for Gaussian distribution whose curve density is represented by equation 3.5 (Thornton et al. 2008).

$$\Delta Z = \sqrt{(\delta Z/\delta X)^2\,(\Delta X)^2 + (\delta Z/\delta Y)^2\,(\Delta Y)^2} \qquad (3.1)$$

$$\Delta Z = \sqrt{(\Delta X)^2 + (\Delta Y)^2} \qquad (3.2)$$

$$\Delta Z = Z\sqrt{(\frac{\Delta A}{A})^2 + (\frac{\Delta B}{B})^2} \qquad (3.3)$$

where X and Y are independent and measurable quantities that are used to obtain a value of a calculated quantity Z; $\delta Z/\delta$ is the partial derivative of the variable Z with respect to an independent parameter (X or Y), and ΔX and ΔY are the uncertainties of the variables X and Y.

$$\Delta\sigma^2 = (\frac{Q\left(m^3/yr\right) \times Z}{1.96})^2 \qquad (3.4)$$

$$F(x; \mu,\sigma^2) = \frac{1}{\sigma\sqrt{2\pi}} e^{-\frac{1}{2}(x-\mu/\sigma)^2} \qquad (3.5)$$

where σ^2 is the variance, Q is the amount of the water balance component (m^3/year), Z is the 95% confidence limit, μ is the mean and σ is the standard deviation.

3.4 TOOLS FOR WATER BALANCE ESTABLISHMENT

Table 3.3 and Table 3.4 show that nine tools are basically water balance tools: AquaLite, AWWA Water Audit, BenchLeak, BenchLoss, CalcuLEAKator, CheckCalcs, Component Analysis, EasyCalc, and WB-PI Calc-UTH. The main focus of these tools is to establish the standard water balance and NRW PIs. Basic system data such as number of service connections, mains, and pressure data are input as well as the water balance data, and the main output is the standard water balance and NRW PIs. However, some tools have more or deeper features. EasyCalc remains the most detailed (Table 3.4), straightforward and comprehensive tool for the water balance establishment. It contains detailed input for system data, pressure data, water balance data, a historical comparison of water balances, and brief what-if scenarios. CheckCalcs, AquaLite, and BenchLoss come next in tolerating essential details about a particular case study. AWWA Water Audit and Component Analysis are tools that are more standardised for water utilities in USA and North American countries, where the input of key figures and water balance components are briefly condensed in a sole or limited input. The limited input has eventually an impact on the sensitivity and the accuracy of the tool. However, the tools (AWWA Water Audit and Component Analysis) have complementary mini-tools for data report, collection and validation. Similarly, WB-PI Calc-UTH and BenchLeak are tools that are locally focused. BenchLeak is one of the first water balance tools and now, is generally outdated and substituted by AquaLite. While all the water balance tools use only the recommended IWA PIs for NRW, WB-PI Calc-UTH has 170 PIs that cover broad aspects of the water service in general. It also has a unique feature of recognising the overbilling practice, which overestimates the billed consumption and subsequently underestimates the NRW. This is the impact of charging a customer a minimum billed consumption (e.g. 10 m^3/month) even though a customer doesn't consume this amount.

Table 3.2. Non-revenue water key performance indicators. Source: Alegre et al. (2016)

| | | | Performance Indicator | | |
| | | | Service Connection Density | | |
Level	Function		> 20/km of mains	< 20/km of mains	Comments
1 - Basic	Financial	NRW	Volume of NRW as % of SIV	Volume of NRW as % of SIV	Simple, not recommended
1 - Basic	Operational	AL	m³/serv. comm./year	m³/km of mains/year	For target setting, not comparing systems
1 - Basic	Operational	RL	L/serv. conn./day	L/km of mains/day	For target setting not comparing systems
1 - Basic	Operational	RL	L/serv. conn./day w.s.p.	L/km of mains/day w.s.p.	Allows for intermittent supply situations
2 - Interim.	Operational	RL	L/serv. conn./day/ m pressure	L/km of mains/day/ m pressure	Useful for comparing systems
3 - Detailed	Financial	NRW	Value of NRW as % of annual cost	Value of NRW as % of annual cost	Allows different unit costs
3 - Detailed	Operational	RL	Infrastructure Leakage Index (ILI)	ILI	Powerful for comparing systems

The impact of the overbilling practice occurs only if the billing system cannot charge the monetary minimum consumption bill unless the volumetric real data in the billing system is altered (manually or automatically) (AWWA 2016). This process is rather destructive for essential costly data of a water utility and causes avoidable opacity of the utility performance. However, when this practice exists in a water utility, the overbilling should be considered in the standard water balance itself and this is the intuition of the WB-PI Calc-UTH tool (Tsitsifli and Kanakoudis 2010).

3.4.1 Intermittency and normalisation

Another key issue is the applicability of these tools in intermittent water supply systems, where water is not available in the network 24/7 and customers adapt to this situation by setting up local storage tanks in their premises. In such a case, the volume of water loss is highly influenced by the volume of the supplied water. The greater is the volume of the water supplied into the network, the more will be the volume of water losses and the higher will be the PIs of NRW, indicating worse performance while it is not necessarily the case (AL-Washali et al. 2019a). To tackle this issue, the volume of water loss and its PIs have to be normalised and adjusted as if the supply is continuous (AL-Washali et al. 2019a). This normalisation process enables monitoring and benchmarking the performance of water loss management in intermittent supply, which is an issue of increasing interest. However, only four tools recognise the intermittency in the tools' input: AquaLite, BenchLeak, CheckCalcs, and EasyCalc, where AquaLite and EasyCalc are relatively more detailed. Although normalising the volume of NRW, AL and its PIs for monitoring and benchmarking is intuitive and critical, it is still a recent highlight that is not considered yet in the four tools. These tools normalise only the RL, explicitly in EasyCalc and AquaLite, and implicitly in CheckCalcs.

3.4.2 Uncertainty analysis

Introducing the uncertainties of the water balance components to the users of the tools is a significant achievement in raising the awareness of the water balance limitations and the practical way to improve them. The uncertainty analysis points out which input data should be more verified in order to minimise the uncertainty of the interesting output. AquaLite, BenchLoss, CalcuLEAKator, and EasyCalc use the variance analysis (Equation 3.4) for identifying the uncertainty of the water loss components. AquaLite has this feature for the water balance components but also for NRW PIs, which is an important gesture of AquaLite. CheckCalcs use the uncertainty analysis, with a similar approach to error propagation theory, to generate uncertainties for identifying the opportunity of pressure management and its influence on reducing the leaks, the bursts, and the consumption (N_1, N_2, and N_3, respectively). It is, however, a key limitation for CheckCalcs and the remaining tools that they don't recognise the uncertainties of the water balance because the water balance is critically influencing all aspects of water loss management intervention and has a great implication on the estimated benefits of each intervention (AL-Washali et al. 2019b). It should be noted that CheckCalcs is just one free tool of many commercial tools that form one package (LeaksSuit) for leakage management.

Relevant to the uncertainty analysis is the use of validation score for the input and output of the tool, to determine the validity and reliability of the tool's output. This is a unique feature of the AWWA Water Audit whose validity score triggers changes in data acquisition rules when a low validity score is recorded. It is equivocal why AWWA Water Audit incorporates the qualitative validity score approach instead of the commonly used uncertainty analysis. Al-Washali et al. (2020b) found that uncertainty analysis helps to improve the outputs of water loss assessment methods, although it did not demonstrate the accuracy or the validity of the methods. Even so, using the uncertainty analysis to improve the output of the tools is not questionable and strongly recommended (Alegre et al. 2016; AWWA 2016; Lambert et al. 2014).

3.4.3 Water loss assessment approach

The approach used to establish the water balance in all the above nine tools, except CalcuLEAKator, is the top-down water audit methodology, where AL are estimated and then RL are calculated. For CalcuLEAKator, the approach used is the MNF analysis (Equations 2.7, 2.8 and 2.9) for one specific DMA and then water balance and NRW PIs are generated for this particular DMA. The tool enables data entry and water balances for 20 DMAs and based on these DMA mini-balances, a global water balance and NRW PIs for the whole network are compiled and created.

As can be noticed from the above description, the top-down water audit is the main approach used to establish the water balance. The MNF analysis and BABE are usually used as complementary analyses for the top-down water audit. However, using more than a method for establishing the water balance for the entire network is recommended (Al-Washali et al. 2020b), because it can improve the accuracy of the tool significantly and assist in establishing more reliable and system-wide balances. For the DMA-scale, MNF analysis remains a powerful methodology to establish the water balance in DMAs.

Table 3.3. Modules and gaps of NRW software tools

#	Model/Tool	Assessment											RL Reduction Planning				AL Reduction Planning		
		NRW	RL	AL	Top.Do. Audit	MNF	BABE	Uncer tainty	Intermi ttency	PIs	Normali sation	Guidance Matrix	PM	ALC	RTM	AM	CMI	DHEs	UC
1	AquaLite v4.5	✓	✓	✓	✓	✗	✗	✓	✓	✓	✓	✗	✗	✗	✗	✗	✗	✗	✗
2	AWWA Water Audit v5	✓	✓	✓	✓	✗	✗	✗	✗	✓	✗	✓	✗	✗	✗	✗	✗	✗	✗
3	BenchLeak	✓	✓	✓	✓	✗	✗	✗	✗	✓	✓	✗	✗	✗	✗	✗	✗	✗	✗
4	BenchLoss (NZ v2a)	✓	✓	✓	✓	✗	✗	✓	✗	✓	✗	✓	✗	✗	✗	✗	✗	✗	✗
5	CalacuLEAKator v4.3	✓	✓	✓	✓	✓	✗	✗	✓	✓	✗	✓	✓	✗	✗	✗	✗	✗	✗
6	CheckCalcs v6b	✓	✓	✓	✓	✗	✗	✗	✗	✓	✓	✓	✓	✗	✗	✗	✗	✗	✗
7	Component Analysis	✓	✓	✓	✓	✗	✓	✗	✗	✓	✗	✗	✓	✓	✗	✗	✗	✗	✗
8	EconoLeak v1a	✗	✓	✗	✗	✗	✓	✗	✗	✗	✗	✗	✗	✓	✓	✗	✗	✗	✗
9	PresMac v4.4	✗	✓	✗	✗	✓	✗	✗	✗	✗	✗	✗	✓	✗	✗	✗	✗	✗	✗
10	SanFlow v4.6	✗	✓	✗	✗	✓	✓	✗	✗	✗	✗	✗	✗	✗	✗	✗	✗	✗	✗
11	WB-EasyCalc v5.16	✓	✓	✓	✓	✗	✗	✓	✓	✓	✓	✓	✓	✗	✗	✗	✗	✗	✗
12	WB-PI Calc-UTH v2.2	✓	✓	✓	✓	✗	✗	✗	✗	✓	✗	✗	✗	✗	✗	✗	✗	✗	✗

Table 3.4. Level of input details of NRW software tools

#	Model/Tool	Level of details				
		SIV	AL	RL	Pressure	Supp. Time
1	AquaLite v4.5	Partial	Partial	None	Full	Full
2	AWWA Water Audit v5	Full	Partial	None	None	None
3	BenchLeak	Partial	None	None	Partial	Partial
4	BenchLoss (NZ v2a)	Full	Partial	None	None	None
5	CalacuLEAKator v4.3	Partial	Partial	Partial	Full	None
6	Check Calcs v6b	Full	Partial	None	Partial	Partial
7	Component Analysis	None	None	Full	Partial	None
8	EconoLeak v1a	None	None	Full	Partial	None
9	PresMac v4.4	None	None	Partial	Full	Partial
10	SanFlow v4.6	None	None	Full	Partial	None
11	WB-EasyCalc v5.16	Partial	Full	None	Full	Full
12	WB-PI Calc-UTH v2.2	Partial	Partial	None	None	None

Level of details:

SIV	Full	Tool considers all components in water supply assessment
	Partial	Tool considers some components in water supply assessment
	None	Tool does not provide details in the assessment of the water supply
AL	Full	Tool considers all sub-components in apparent loss assessment with options
	Partial	Tool considers some sub- components in apparent loss assessment
	None	Tool does not provide details in the assessment of the apparent loss
RL	Full	Tool considers all sub-components in real loss assessment
	Partial	Tool considers some sub-components in real loss assessment
	None	Tool does not provide details in the assessment of the real loss
P	Full	Tool considers the average pressure in each zone
	Partial	Tool considers the overall average pressure
	None	Tool does not provide details in the pressure assessment
T	Full	Tool considers the supply time per hour for each zone
	Partial	Tool considers the supply time as percentage of time pressurised for whole system
	None	Tool does not consider the intermittency and duration of the supply in the assessment

3.5 TOOLS FOR WATER LOSS REDUCTION PLANNING

Out of the 12 available tools, many tools touch on several aspects of planning for water loss reduction for the whole network: five tools provide guidance matrix for leakage reduction intervention, two tools accommodate economic analysis, and three tools indicate the opportunity of global pressure management.

3.5.1 Guidance matrices

The common guidance matrix is shown in Table 3.5. The matrix was developed for the World Bank Institute as a target matrix and a banding system for leakage performance categories. The limits of categories for low and mid-income countries were set as twice the allowance of high-income countries (Lambert 2015a), to set feasible targets for water utilities in low and mid-income countries. Having the volumetric leakage level or through the ILI, the leakage category of a certain utility is easily defined in Table 3.5. Based on the categories A1, A2, B, C, and D, different recommendations are provided (Liemberger and Partners 2018):

A1: small potential for further NRW reductions; A2: further NRW reduction may be uneconomic unless there are water shortages or very high water tariffs; B: potential for marked improvements; establish a water balance, consider pressure management, active leakage control, better network maintenance, improve customer meter management, review meter reading, data handling and billing processed and identify improvement potentials; C: poor NRW record; tolerable only if water is plentiful and cheap; even then, analyse level and causes of NRW and intensify NRW reduction efforts; and D: highly inefficient; a comprehensive NRW reduction program is imperative and high-priority.

Table 3.5. Leakage assessment matrix. Source: EasyCalc v5.16

| | | | | Leakage (Litres/connection/day) with P_{avg}: | | | | |
		ILI	10 m	20 m	30 m	40 m	50 m
	A1	< 1.5		< 25	< 40	< 50	< 60
	A2	1.5 - 2		25-50	40-75	50-100	60-125
Standard	B	2 - 4		50-100	75-150	100-200	125-250
	C	4 - 8		100-200	150-300	200-400	250-500
	D	> 8		> 200	> 300	> 400	> 500
	A1	< 2	< 25	< 50	< 75	< 100	< 125
Low and	A2	2-4	25-50	50-100	75-150	100-200	125-250
Middle Income	B	4 - 8	50-100	100-200	150-300	200-400	250-500
Countries	C	8 - 16	100-200	200-400	300-600	400-800	500-1000
	D	> 16	> 200	> 400	> 600	> 800	> 1000

The matrix presented in Table 3.5 is provided in EasyCalc, BenchLoss and CalcuLEAKator. EasyCalc provides this matrix but also another similar matrix for the total volume of NRW. CheckCalcs has a similar leakage matrix but with splitting the B,

C and D categories into two sub-categories for each category, following practices in Australia and Malaysia, with exactly the same approach of A1 and A2 in Table 3.5. AWWA Water Audit has, in turn, pre-set targets for the ILI and coupled with technical and financial considerations. These matrices are useful and commonly applied, however, they are developed based on mere experience, not a scientific foundation nor published materials with deeply-studied data.

3.5.2 Economic leakage detection

EconoLeak estimates the total volume of the leakage, background losses and their costs. Afterwards, it aggregates the leakage reduction cost in terms of sounding cost, leak correlation cost, MNF cost, repair cost, and finally administration and supervision cost. A simple cost-benefit analysis enables plotting of the curve of the short-run economic level of leakage or the economic leakage detection. Figure 3.1 represents the main output of EconoLeak, where the x-axis represents the leakage level and also the number of days required to survey the whole network using the leakage detection techniques. The active leakage control curve (green curve) shows that the detection survey becomes dramatically costly when the survey period of the whole system is less than one year and becomes more economic when it is more than a year (Mckenzie and Lambert 2002). The lowest point in the total cost curve (pink curve) in Figure 3.1 corresponds to the economic survey period, which is almost annually in this example. The dotted vertical lines represent the base, economic and unavoidable levels of leakage of this example system. Any further leakage reduction after the economic (accepted) threshold in Figure 3.1 becomes basically uneconomic.

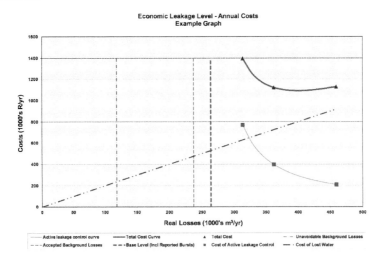

Figure 3.1. Economic level of leakage detection. Source: EconoLeak software v1.a

Obviously, this approach requires extensive cost data for all single elements of intervention. The tool in fact does not create the (long-run) economic level of the leakage but only the economic intervention frequency of the leakage detection survey. It does not consider other interventions such as pressure management, assets management and response time minimisation. The economic level of the leakage should be determined based on all possible intervention for a specific case and not only the leakage detection survey. Alternatively, the tool Component Analysis has a more matured feature of defining the economic intervention frequency. It defines how frequent should the leakage detection survey complete the entire network. This is estimated based on the variable cost of water, intervention cost, and also the rate of rise of unreported leakage, as shown in Equations 3.6 and 3.7 (Lambert and Fantozzi 2005).

$$EIF = \sqrt{0.789 \times \frac{CI}{CV \times RR}} \qquad (3.6)$$

$$EP = 100 \times 12/EIF \qquad (3.7)$$

where EIF (months) is the economic intervention frequency through the leakage detection surveys, CI is intervention cost (\$/Km), CV is variable cost (\$/m^3), RR is the rate of rise of unreported leakage (m^3/Km mains/day/year), and EP is Economic Percentage of system to be surveyed annually. The limitation of Equation 3.6 is the variable RR, which is sensitive and difficult to estimate. However, guidance for estimating this factor is available in Fanner and Lambert (2009). Another concern of Equation 3.6 is its probable overestimation of the leakage detection potential, as discussed in AL-Washali et al. (2019b).

Interestingly, the tool Component Analysis does not only estimates the frequency of the leakage detection survey, but also gives an opportunity in doing cost-benefit analysis of pressure management and response time minimisation for the whole network. Firstly, the tool enables setting the exponent of the pressure-leakage relationship (N$_1$; Equation 3.8) and then the tool estimates the potential volumetric and monetary savings of pressure reduction. It is worth mentioning that the leakage discharge from pressurised pipes varies with pressure during the day as highlighted by the concept Fixed and Variable Area Discharges (FAVAD) (Lambert 2001; May 1994), which overcomes the limitations of Torricelli equation for plastic pipes. Equation 3.8 presents the leakage-pressure relationship that is reconciled with intensive empirical research, using Japanese and UK data. The leakage exponent N$_1$ in Equation 3.8 varies from 0.5 for rigid pipe materials to 1.5 for variable leakage area in flexible (plastic) pipe material, whose leaks' split varies with pressure.

$$\frac{L_1}{L_0} = \left(\frac{P_1}{P_0}\right)^{N_1} \qquad\qquad (3.8)$$

where L is the leakage volume, P is the pressure and N_1 is the leakage-pressure exponent. Similarly, the Component Analysis tool estimates the potential volumetric and monetary saving of minimising the repair-response time, based on the direct cut of the leaks' run-time. Expectedly, the tool does not recognise the economic aspect of assets management, perhaps because estimating the benefits of the assets management is complex and it is typically not cost-effective. This tool is probably the most economically comprehensive tool for leakage management that is freely available. Example of the output of this tool is presented in Al-Washali et al. (2020b) and AL-Washali et al. (2019b).

3.5.3 Global pressure management opportunity

There are many network simulation and hydraulic modelling tools that consider pressure management and pressure control for water networks. This is basically based on the concept that optimising the pressure in the network triggers significant leakage reduction. Incorporating pressure management with DMA demarcation is commonly utilised by such tools. An example of these tools is the freely available tool, EPANET and its possible add-ins modules. Several (commercial) tools are also available, including: WaterGYMS, InfoWater, WDNetXL, H_2O MAPWater, KYPIPE and other tools. However, there are three freely available tools that snapshot the pressure management opportunity for the whole (global) network and estimate the feasibility of pressure management without the need of hydraulic simulation of the network pipes and appurtenances, which is the focus of this chapter. These tools are: CheckCalcs, Component Analysis, and EasyCalc. The principle is similar and intuitive. Based on defining the exponent of the leakage-pressure relationship (N_1) in Equation 3.8, the reduction on the leakage volume as a consequence of reduced pressure can be estimated, and then its monetary value can be easily derived. This is what EasyCalc exactly does. However, CheckCalcs further analyses the impact of pressure reduction on the reduction of bursts frequencies as well as the customers' consumption (N_2 and N_3 respectively). The Component Analysis, however, enables estimating the benefit of pressure management and comparing it to its cost as well as the costs and benefits of other leakage interventions.

3.6 TOOLS FOR WATER LOSS MANAGEMENT INTERVENTION

The previous two sections tackle the assessment of water losses and planning for its management. However, when it comes to working on the ground during the intervention phase, only the tools that are focused on DMA-scale are hands-on. These tools are SanFlow, and PresMac.

3.6.1 Active leakage detection

Similar to CalcuLEAKator which is discussed in section 3.4, SanFlow is a powerful tool that also analyses MNF in a DMA within the network using Equations 2.7, 2.8 and 2.9. It is a suitable and operational tool to intervene and reduce the leaks in a DMA. It estimates the legitimate night use, the leakage and the unavoidable background losses using the BABE approach. After deducing the unavoidable background losses, the tool transforms the leakage volume for each DMA into an estimated equivalent number of bursts in the DMA. Notably, the tool does not assume the presence of bursts within each DMA, but use this as an index to rank the DMAs based on their leakage level. Comparing this equivalent number of bursts for all the DMAs assists in prioritising the DMAs for leakage minimisation interventions. SanFlow estimates the hourly leakage only during the MNF time and does not provide estimates on a daily leakage level, let alone annually. Figure 3.2 gives an insight into the output of SanFlow. Figure 3.2 implies that there was significant leakage in the DMA at the beginning (the yellow area) and after some time, when the leaks were fixed, the pressure in the DMA enhanced, causing more unavoidable background leakage in the DMA that cannot be sensed by the available leakage detection technology. So the blue line, which shows the estimated equivalent service pipe bursts, represents the priority level of intervention in this DMA, and can be compared with other DMAs in the network which is a main function in this tool.

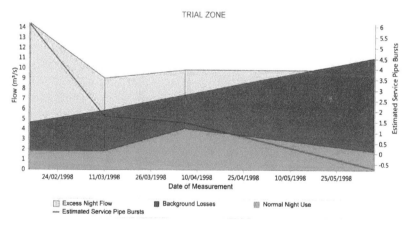

Figure 3.2. Leakage, background losses and night use in an example DMA. Source: SanFlow v4.6

3.6.2 Zonal pressure management

The potential for pressure management is consistently promising. Reducing the pressure at a critical point in a DMA by a certain level triggers a more pressure reduction at the inlet of the DMA (Mckenzie and Langenhoven 2001). PresMac, therefore, assesses the monetary savings of pressure reduction due to installing fixed-outlet and time-modulated PRVs in a specific DMA. The tool does not rely on hydraulic representation of the DMA's network, rather, it compromises MNF analysis, BABE analysis, N_1 exponent estimation, and further analysing pressure measurements, friction factors (K) and head losses (H_L) in three key points in the DMA: the inlet point, the average (elevation) zone point, and the critical point(s) in the DMA whose pressure is the lowest in the DMA during the course of the day. The fixed-outlet PRV dictates a fixed pressure during all the hours of the day (Figure 3.3-b). The time-modulated controller reduces the outlet pressure at certain times of the day when the demand is basically reduced (Figure 3.3-c). In a critical point modulation, the pressure is sensed at the critical point and communicated to the PRV at the inlet which, in turn, adjusts the inlet pressure to maintain the minimum pressure at the critical point during the course of the day. This process is effective, but further investment is required. In PresMac, the flow-modulated controller (Figure 3.3-d) dictates the inlet pressure in accordance with the instantaneous demand and the excessive pressure at the critical point in the DMA. While firefighting flows cannot be met using the time-modulated controller, it can be satisfied using the flow-modulated controller. The use of PRVs optimises the pressure in the DMA and achieve the minimum leakage. PresMac, however, analyses only the benefits of fixed-outlet and time-modulated PRVs. If the less expensive time-modulated controller is economically justified, the flow-modulated controller can provide greater benefits. A limitation of pressure reduction is the potentially small reduction in customer demand. In this regard, PresMac assumes that leaks are pressure-dependent and consumption is pressure–independent, although some consumption is in fact pressure-dependent, such as washing hands, brushing teeth and garden irrigation.

The approach used in PresMac is iterative. Firstly, using MNF measurements and BABE analysis, the tool estimates the background leaks and solves Equation 3.8 to estimates the value of N_1 as shown in Equation 3.9. Secondly, the tool splits the inflow into the DMA into pressure-dependent and pressure-independent flows. Then, the tool estimates the friction factor (K) using Equation 3.10 for each hour of the day at two points: the average zone point and the critical point in the DMA.

$$N_1 = \frac{log(\frac{L_0}{L_1})}{log(\frac{P_0}{P_1})} \qquad (3.9)$$

$$K = \frac{H_L}{Q^2} \qquad (3.10)$$

where K is head loss coefficient ($m^{-5}.h^2$), H_L is head loss in m, and Q is flow in (m^3/h).

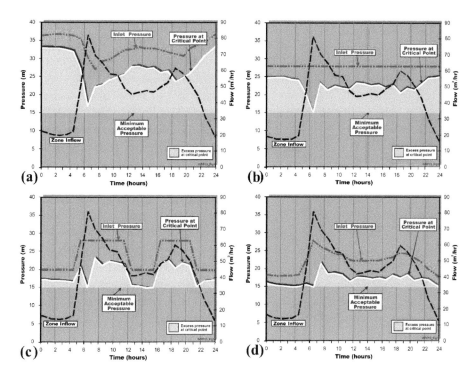

Figure 3.3. Pressures at a critical point in a DMA: (a) no PRV, (B) a fixed-outlet PRV, (c) a time-modulated PRV, and (d) a flow-modulated PRV. Source: manual of PresMac v4.4

The tool then selects a fixed outlet pressure and recalculates pressure in the average zone point and the critical point using K factors and the corresponding zone inflow. This process is iterative until the minimum acceptable pressure at the critical point is achieved. This process is carried out for the fixed-outlet PRV which saves significant leakage volume. Nonetheless, the achieved minimum acceptable pressure at the critical point is only for the peak hour and is still higher than the required pressure during the remaining hours of the day, when the demand is lower. Therefore, the time-modulated controller can further reduce the leakage by changing the times and switching from high to low-pressure periods during the day, and the benefits of this option are particularly analysed in PresMac. The benefits of adopting flow-modulated PRV in the DMA are unfortunately not provided in PresMac. Finally, it is worth mentioning that recent researches on N_1 exponent (Lambert et al. 2017a; Van Zyl and Cassa 2014) revealed that N_1 itself is affected by the changing pressure during the day and this particular issue is still not considered in PresMac.

3.7 SUMMARY OF THE TOOLS' MODULES AND GAPS

Table 3.3 summarises the modules and gaps for each individual software tool. The functions of these tools fall into the above three categories: (i) water balance establishment; (ii) water loss reduction planning; and (iii) water loss management intervention. Although asset management is a main pillar of RL reduction, it is still not recognised in the NRW reduction tools. Besides, while AL is a major concern in low- and mid-income countries, it is not sufficiently recognised in the input data for most of the tools. Providing more detailed input for AL results in more accurate results. Table 3.4 shows the level of details of the tools in terms of AL and other key parameters that affect the accuracy of the water balance. Table 3.4 shows that only one tool, the WB-EasyCalc, has a sufficiently detailed module for AL. Regarding AL reduction, Table 3.3 shows that AL reduction planning has garnered relatively little attention among software developers, which is a major limitation in the industry in general. A good start in this regard would consist of developing a tool or a module to optimise policies for customer meter replacement in the distribution network. It is worth to mention that the reviewed tools are mainly developed by practitioners in the field, aiming to be applicable in real-world and for different network sizes. The reviewed tools use hydraulic analysis and hydraulic modelling to a certain extent, however, they do not incorporate hydraulic representation of the network's pipes and appurtenances neither for the whole network nor for a DMA in the network.

3.8 GUIDANCE FOR THE USE OF NRW SOFTWARE TOOLS

Figure 3.4 presents a fit-for-purpose guide for selecting an NRW software tool to establish the water balance and generate NRW PIs. When using the top-down water balance, if the water supply is intermittent, then three tools can be used: WB-EasyCalc, Check-Calc and Aqualite. If the water supply is continuous and overbilling of consumption is of concern, then WB-PI Calc-UTH can be used. Otherwise, Water Audit, WB-EasyCalc, Check-Calc, Aqualite, and BenchLoss can be used. When the water balance needs to be established using MNF, then CalcuLEAKator can be used.

Similarly, Figure 3.5 presents a fit-for-purpose guide for selecting an NRW software tool when assessing water loss to plan for reduction measures. Planning for RL reduction requires conducting a BABE analysis. Planning for RL reduction can be accomplished using two scales: the whole network or within a DMA. For the whole network, a BABE analysis can be conducted using the tool Component Analysis. To analyse the potential for leakage reduction when carrying out pressure management interventions, the tools Component Analysis, Check-Calc, and WB-EasyCalc can be used.

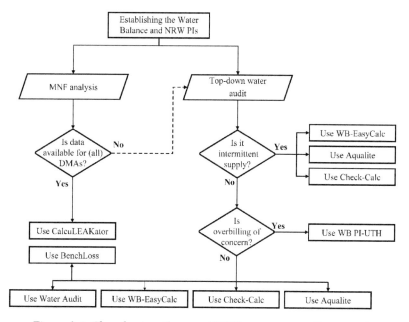

Figure 3.4. Flowchart on the use of NRW assessment software tools

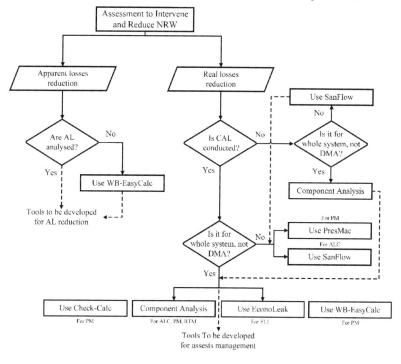

Figure 3.5. Flowchart on the use of NRW intervention software tools

To analyse the economic level of leakage, EconoLeak can be used. To analyse the potential for leakage reduction when minimising the response times for reported bursts or when carrying out active leakage detection, the Component Analysis tool can be used. To plan for RL reduction on a DMA-scale, Sanflow is first used to conduct a BABE analysis. Afterward, Sanflow is again used to prioritise between DMAs for active leakage detection and then PresMac can be used for pressure management analysis. Unfortunately, planning for AL reduction is still missing and tools need to be developed for this purpose.

3.9 CONCLUSIONS AND FUTURE PROSPECTS

The availability of several NRW software tools and many iterations of each individual tool indicate the promising future of NRW software development. This chapter reviewed 12 freely available NRW software tools for assessing NRW in distribution networks. The tools use hydraulic analysis in their modules, but they do not hydraulically model a representation of the network's pipes and appurtenances. Most (9) of the tools have been developed to establish the standard annual water balance and NRW recommended performance indicators for the whole network. Some (3) tools, however, have been developed to intervene and reduce leakage within a DMA.

The main aspects of the software tools that are currently improving or need to be improved further in the software tools are: uncertainty analysis, consideration of supply intermittency, and loss reduction analyses. Although the importance of uncertainty analysis is widely recognised in the industry, only 5 tools have a module for uncertainty analysis. Intermittency of the water supply is a subject of increasing interest, yet it is considered in only 4 tools. Whenever intermittency is of concern, normalising the PIs to a continuous supply is necessary, and this has been included in only 4 tools (which consider only real losses). Normalising the NRW and apparent losses remain unaddressed. To plan for leakage reduction, the tool Component Analysis model is relatively comprehensive as it analyses the potential benefits of pressure management, active leakage detection, and minimising response and repair time of bursts. However, apparent losses have still not been adequately recognised in the industry.

Although a comprehensive NRW management tool for monitoring, planning, and intervention is currently not available, such a tool can be developed in one complete package or in a kit of several tools. The modules presented in Table 3.3 should be included in such a tool. However, certain critical aspects should be emphasised. Firstly, recognising the intermittency of the supply will widen the use of NRW software tools. For this normalising all the volumes and performance indicators of the NRW, apparent losses, and real losses is essential. Currently, normalisation is only considered for real losses and not for apparent losses and NRW. Secondly, distinguishing the scales of the tool is crucial. Planning for loss reduction should first be performed for the whole network;

then planning for specific interventions should be done on a DMA-scale. In both scales, a comprehensive tool that considers different reduction measures for both real and apparent losses should be useful and effective. Thirdly, greater focus needs to be placed on the apparent loss estimation and minimisation; the reduction of AL is cost-effective and it is relevant to, and considered a priority for, many water utilities in low- and mid-income countries. Fourthly, a comprehensive tool for NRW would definitely benefit from including capabilities for uncertainty analysis, guidance matrix and assessing NRW components using different methods. Finally, NRW software tools should include the capability to determine the economic level of water loss based on estimating and combining the economic levels of leakage and apparent losses.

3.10   SUPPLEMENTARY INFORMATION ON THE NRW ASSESSMENT TOOLS

Aqualite

Aqualite is a Windows-based tool developed by Mckenzie (2007) and the Water Research Commission (WRC) in South Africa to establish the top-down water balance and NRW PIs. It is an updated version of the tool BenchLeak. Aqualite can use different units and the intermittency of the supply and uncertainty analysis are included in the tool. However Aqualite normalises only the PIs of the RL, assuming that the supply is continuous. Volumes and PIs of NRW and AL remain unnormalised and are therefore affected by the level of intermittency and the changes in the duration of the supply.

AWWA Water Audit

Water Audit is an Excel-based tool developed by the American Water Works Association (AWWA) (Water Loss Control Committee 2014) to establish the water balance and NRW PIs. A unique feature of this tool is the use of validity scores and guidance instead of uncertainty analysis. The validity guidance triggers changes in data acquisition rules when a low validity score is recorded. However, it is questionable why the tool uses the qualitative validity score approach instead of the commonly used uncertainty analysis. Al-Washali et al. (2020b) found that uncertainty analysis helps to improve the outputs of water loss assessments, although it did not demonstrate the accuracy or the validity of the method. Another feature of this tool is its use of a guidance matrix for the input data to plan water loss control. However, this tool does not consider the intermittency of the supply and therefore all NRW PIs are not normalised. Moreover, it provides limited details for the AL, i.e. it uses one figure for each component of AL, without providing for the possibility of more details.

BenchLeak

This is an Excel-based tool developed by the WRC to establish a water balance and compute NRW PIs (Mckenzie et al. 2002). Its functionalities are very basic and as an initial tool for analysis, has been substituted by AquaLite. However, BenchLeak remains freely available for generating the NRW PIs for water utilities in South Africa. A good feature of this tool is the associated user manual which clarifies the concepts behind the calculations made in the tool. However, BenchLeak is, in general, outdated.

BenchLoss

BenchLoss NZ is an Excel-based tool developed by Global Water Resources (GWR) for water utilities in New Zealand (GWR-Ltd 2008). The tool assists in establishing the water balance and NRW PIs and benchmarks the performance relative to other utilities in New Zealand. BenchLoss intensively explains the input and the output of the tool and identifies appropriate action priorities based on a guidance matrix. The tool estimates the confidence limits of the inputs and the uncertainties of the outputs. It also assesses the details of AL components. However, BenchLoss does not consider the intermittency, normalisation, and other features of leakage reduction.

CalcuLEAKator

This is an Excel-based tool designed by an independent consultant (Koldžo and Vucˇijak 2013) to analyse and compile MNF analyses. The tool can generate the water balance and NRW PIs based on both top-down and bottom up approaches, which is a unique feature of the tool. However, establishing the water balance for the whole network based on MNF data or measurements of some DMAs is always questionable, because MNF analysis in one or several DMAs during a specific period will always be different from other parts of the network and at other times of the year.

Component Analysis Model

Component Analysis is a more comprehensive Excel-based model developed by the AWWA Water Research Foundation (WRF) (Sturm et al. 2014). It establishes the water balance and NRW PIs and analyses the components of the RL using the BABE method for the whole network. It then analyses the potential and the economic feasibility of reducing the leakage through pressure management (PM), active leakage detection and control (ALC), and repair response time minimisation (RTM). Component Analysis was developed for water utilities in North America and therefore, has a benchmarking feature to other utilities in North America. However, this tool does not consider intermittency, uncertainties, and normalisation.

EconoLeak

This is an Excel-based tool developed by the WRC for determining the economic level of leakage (ELL) (Mckenzie and Lambert 2002) by plotting the curve of the ELL based on cost and benefit estimation of ALC in the network. The tool has a user guide explaining the principles behind the calculations of the model. However, the tool was developed for water utilities in South Africa and therefore uses local currency and cost estimates.

CheckCalcs

CheckCalcs is an Excel-based tool developed by ILMSS Ltd for establishing the water balance and NRW PIs (Lambert 2015a). It includes detailed instructions, it benchmarks the system being analysed to 12 other European systems; and it provides a guidance matrix to identify appropriate action priorities. In addition, this tool gives insights on the probable changes in leaks, bursts, and consumption (N_1, N_2, and N_3, respectively) when the pressure of the system is changed. However, this model accepts limited details and inputs for determining the AL.

PresMac

PresMac is a Windows-based tool developed by the WRC as an operational tool for pressure management in a DMA (Mckenzie and Langenhoven 2001). PresMac includes a guidance manual explaining the concepts of the tool. The tool features fixed outlet as well as time-modulated analysis for pressure reducer valves (PRVs); it is the only free pressure management tool that works on a DMA scale. PresMac considers both the pressure-dependent and -independent flows for leaks and for legitimate night uses. A unique feature of this tool is its ability to calculate the value of the pressure-leakage relationship exponent (N_1) based on night-time inflow and pressures within a DMA. While estimating the relationship between the leakage exponent N_1 and the fluctuating pressure in a DMA during the day is a matter of increasing concern (Lambert et al. 2017a; Van Zyl and Cassa 2014), this is not considered in the model.

SanFlow

SanFlow is a Windows-based tool developed by the WRC to analyse night flows using the MNF analysis approach in a DMA (Mckenzie 1999). It also breaks down the RL in the DMA into its component parts using the BABE method. However, SanFlow does not calculate the uncertainties associated with the outputs. It also estimates the hourly leakage only during the MNF time and does not provide estimates on a daily leakage level. Instead, the tool transforms the leakage volume during the MNF hours into equivalent estimated bursts. Comparing this equivalent number of bursts for all the DMAs assists in prioritising the DMAs for leakage minimisation interventions.

WB-EasyCalc

WB-EasyCalc is an Excel-based tool developed by Liemberger and Partners (2018) to establish water balance and NRW PIs. This tool has many advantages. It is tidy and user-friendly; it provides details for all the components of the water balance and considers the uncertainty of the outputs. It also recognises intermittency of supply and therefore has normalised PIs for the RL. However, the volumes of the NRW and AL and their PIs are unnormalised in this tool. The tool features a ''what if'' scenario analysis when changing the pressure or the duration of the supply in the network. It also enables historic water balance data comparison. However, the tool would definitely benefit from the inclusion of the component analysis of the RL and consideration of the overbilling practices in the system.

WB-PI Calc-UTH

This is an Excel-based tool developed by Tsitsifli and Kanakoudis (2010) to establish water balance and NRW PIs. This is the only tool that considers overbilling practices in the water balance. However, the tool is in Greek; and while it considers 170 PIs that may be suitable for the local context, these are not necessarily suitable for other utilities.

4

MONITORING NON-REVENUE WATER IN INTERMITTENT SUPPLY

Water utilities should monitor their non-revenue water (NRW) levels properly to manage water losses and sustain water services. However, monitoring NRW is problematic in an intermittent water supply regime. This is because more supplied water to users imposes higher volumes of NRW, and supplying significantly less water results in an unmet water demand but interestingly less NRW. This study investigates the influence of the amount of water supplied to a distribution system on the reported level of NRW. The volume and indicators of NRW all vary with variations in the system input volume (SIV). This is even more critical for monitoring NRW for systems shifting from intermittent to continuous supply. To enable meaningful monitoring, the NRW volume should be normalised. Addressing that, this chapter proposes two normalisation approaches: regression analysis and average supply time adjustment. Analysis of the NRW performance indicators showed that regression analysis enables the monitoring of NRW and tracking its progression in an individual system, but not for a comparison with other systems. For comparing (or benchmarking) a water system to other systems with different supply patterns, the average supply time adjustment should be used. However, this approach presents significant uncertainties when the average supply time is less than eight hours per day.

This chapter has been published as: AL-Washali, T., Sharma, S., AL-Nozaily, F., Haidera, M., and Kennedy, M. "Monitoring the Non-Revenue Water Performance in Intermittent Supply." Water, 11(6), 1220, 2019.

4.1 INTRODUCTION

Non-revenue water (NRW) either originates from leakages that occur at mains, storage reservoirs, and customer connections, or apparent losses that occur due to customer meter under-registration, errors in data handling and billing, or unauthorised use. The impact of NRW is substantial (Erickson et al. 2017; Pillot et al. 2016), as it accounts for considerable water wastage, affects the technical stability of the water supply, deteriorates the quality of water and water services, increases the operating costs, and reduces revenues that should sustain and expand access to water. The level of NRW is also considered as an indicator of the operational efficiency of the water system, water utility governance, and the physical condition of the water supply system (Male et al. 1985; McIntosh 2003; Park 2007; Wallace 1987). The level of leakage is likely the most important single indicator of the efficiency of water utilities perceived by regulators, the public, and the media (European Commission 2015). Utilities and projects are typically evaluated based on pre-set criteria, among which the NRW level is important (KFW 2008; Van den Berg and Danilenko 2010).

To help water utilities sustain their services and manage their losses, NRW levels should be properly monitored. This is a crucial step for effective water loss management (AL-Washali et al. 2016; Mutikanga 2012), and becomes more critical for utilities with variations in water productions and supply hours. When the amounts of water produced and distributed are higher, water typically remains in networks for a longer time, presenting challenges such as longer leakages and more potential for theft and other apparent losses (AL). Similarly, when less water is supplied, the NRW volume will be lower and NRW performance indicators (PIs) imply more efficient performance, but this may not be the case. Although this issue appears critical and intuitive, it has not been recognised in the literature (Al-Ansari et al. 2014; Alegre et al. 2016; Carpenter et al. 2003; Cunha Marques and Monteiro 2003; Ermini et al. 2015; Korkmaz and Avci 2012; Mutikanga et al. 2010; Pybus and Schoeman 2001; Renaud et al. 2014; Staben et al. 2010). The influence of varying amount of supplied water on the amount of NRW trigger the question what should be monitored, the volume of NRW or the management status of NRW. For intermittent supply, the aim should be to improve the service, increase the water production and move towards a continuous supply. However, in this context, it is challenging to monitor the progress of NRW management and intervention, as all PIs vary according to variations of supplied water.

This chapter elucidates the relationship between NRW and system input volume (SIV), and analyses the potential methodologies for monitoring NRW and assessing its performance in an intermittent water supply regime. This analysis can help water utilities to properly monitor NRW levels and alleviate the political pressure that may be subjecting NRW management to failure in situations where certain measures are not at fault, such as systems with increasing levels of supplied water or supply times. It also reveals the reality of poor performance when the NRW level is lowered due to a decrease in the amount of water supplied to customers, rather than the implementation of effective reduction measures. The chapter first analyses the influence of changing the SIV by varying the supply time when the network's pressure is constant on the volume and PIs of NRW. Then it discusses the normalisation of the NRW volume and PIs through "when-system-is-pressurised" (w.s.p.) adjustment and regression analysis, using the city of Sana'a, Yemen, as a case study. Monitoring NRW management using the suggested normalisation approach is the way forward for NRW management in intermittent supply. The chapter provides an overview of the case study and then compares NRW PIs with and without normalisation, highlighting the appropriate methodology of NRW management monitoring.

4.2 RESEARCH METHODOLOGY

4.2.1 Overview of Sana'a water supply

Sana'a City is the capital of Yemen, a water-scarce country in the Middle East. The water situation in the Sana'a Basin is severe, as the water table drops by 6-8 m annually (JICA 2007). The public water service of Sana'a provides water to 45% of the city's population, and the remainder receives water from the private sector, mainly via water tankers. The utility serves 94,723 customers, constituting approximately 1.5 million consumers. The only source of water is a deep aquifer, from which water is extracted at 114 wells with depths ranging from 600 to 1000 m below ground. The water supply in Sana'a is a combined system, employing both pumped and gravity supply, and approximately 50% of the network is pumped from the main headworks. The supply network is geographically divided into six administrative zones, and each zone is subdivided into distribution areas, in which a total of 369 areas are interlinked and multi-fed. The supply pattern in Sana'a is intermittent and insufficient, as it does not fully meet the customers' demands. A customer received water once a week, with an average supply time of 4.4 hr/day, prior to the current severe situation in Yemen that began in March 2015 (AL-Washali et al. 2018). The shortage of water necessitates a rationing program, and the implementation of an intermittent supply has caused network deterioration, accompanied by water quality deterioration and inadequate pressure in sections of the network.

Figure 4.1 (continuous line) presents the volume and apparent trend of NRW based on data from Sana'a water utility for 2005-2015. The average level of NRW in Sana'a for the 10 years was 7.1 million cubic metres (MCM) (35% of SIV), without considering inaccuracies of the production meters. Since 2011, Yemen has been facing great instability, with fuel and electricity shortages, that has further impacted the water services in Sana'a. This situation became critical in 2015, when a Saudi-led military campaign began in Yemen. Consequently, the energy sources for water production from wells have further declined, reducing the amount of water supplied to customers. Accordingly, the NRW level was reported to have decreased significantly to 2.8 MCM (22% of SIV) in 2015 (Figure 4.1). The dotted line in Figure 4.1 shows the normalised volume of NRW in Sana'a, which shows the NRW status in Sana'a (2011-2015) as discussed in details in the following sections.

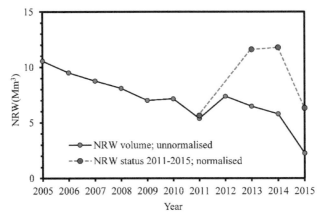

Figure 4.1. NRW volume (unnormalised, apparent trend) and NRW status (normalised trend) in Sana'a.

4.2.2 Analysis of NRW and SIV trends

The trends of SIV, billed consumption, and NRW were analysed over 10 years. Monthly measurements of SIV and billed consumption were obtained from the Sana'a water utility and then converted to annual volumes. The monthly and annual volumes of NRW were then calculated using Equation 2.1 (Alegre et al. 2016; Vermersch et al. 2016). The trends of the volumes of NRW, SIV, and billed consumption were analysed monthly and yearly. As NRW reduction activities are a combination of top-down and bottom-up approaches, the analysis was conducted system-wide and at district metered area (Farley and Trow 2003) scales to investigate the agreement between the results obtained at both scales.

4.2.3 NRW component assessment

The volumes of NRW components were calculated using Equations 2.1 and 2.2. The volume of unbilled authorised consumption (UAC) was estimated by auditing and analysing the records of the Sana'a water utility. Unbilled authorised consumption has two types; metered and unmetered. The unbilled metered consumption in Sana'a water utility consists of (i) the consumption of the staff in the utility's metered buildings; and (ii) the exemptions and wells guards' consumption. The unbilled unmetered consumption consists of water used for washing the pipelines, firefighting, special institutions, and consumption of some notable and powerful people. All these unmetered categories are supplied by means of water tankers and thus estimated by the number of the tankers per year for each category in every administrative zone. The unbilled authorised consumption (metered and unmetered) was calculated by obtaining relevant data from the records of the utility.

The volume of the apparent losses (AL) was estimated using the apparent losses estimation equation, as elaborated by AL-Washali et al. (2018). The volumes of real losses (RL) and water loss were calculated using Equations 2.1 and 2.2. Accordingly, the International Water Association (IWA) standard water balance was established for the Sana'a water supply system for 2009. As the data of 2015 are incomplete, the same estimated proportions of NRW components in 2009 were used to calculate the NRW components for the year 2015, and the results were compared. The uncertainties of the NRW components were analysed using the error propagation theory (Taylor 1997). The error theory equations used are elaborated in Chapter 3.

4.2.4 NRW performance indicators

Liemberger and Farley (2004) and Alegre et al. (2016) recommended the IWA best practice PIs for NRW management, as presented in Table 3.2 in Chapter 3. Frauendorfer and Liemberger (2010) recommended the w.s.p. approach to normalise the overall level of NRW. In this approach NRW is divided by the average supply time of the system, when the system is operated and water is supplied (h/day) and then multiplied by 24 (h/day). Because in an intermittent supply the average supply time is always less than 24 h/day, w.s.p. adjustment should in principle increase the NRW volume assuming the supply is continuous. For leakage monitoring and benchmarking, the European Commission (2015) categorised two fit-for-purpose key performance indicators (KPIs) for targets and tracking progress in individual systems, including volume/year, m^3/km of mains/day, litres/connection/day, and litres/property/day. The KPIs for comparing internal/external leakage between different systems are the Unavoidable Annual Real Losses (UARL), Infrastructure Leakage Index (ILI), average pressure, value of leakage Euro/m^3, and repair frequencies. The infrastructure leakage index was calculated using Equation 4.1 (Lambert et al. 2014).

$$ILI = \frac{CARL}{UARL} \qquad (4.1)$$

where CARL is the current annual RL and UARL is the unavoidable annual RL, which can be calculated from Equation 2.12 in Chapter 2. Vermersch et al. (2016) suggested that the Apparent Loss Index could be used, which can be calculated in a similar manner to the infrastructure leakage index using Equation 4.2.

$$ALI = \frac{CAAL}{RAAL} \qquad (4.2)$$

where CAAL is the current annual AL and RAAL is the reference annual AL, which represent 5% of the volume of the billed, authorised metered consumption, excluding exported water.

Suitable KPIs are essential for the effective monitoring and management of water loss. The package of NRW PIs should be designed as a "fit-for-purpose" set of indicators. However, they should be clearly defined, auditable, quantifiable, achievable, and interpretable (European Commission 2015). For Sana'a, the recommended, best-practice volume indicators of NRW and its components were calculated and adjusted to per-connection indicators as the service connection density in Sana'a is 97 service connections per km of mains. The per-pressure RL indicators and AL index were also calculated and analysed.

4.2.5 Normalising the NRW PIs using w.s.p. adjustment

'When-system-is-pressurised' adjustment is often used to normalise the PIs of the RL of systems with intermittent supply to allow comparison with other continuous supply systems. The volume of losses is adjusted as though the supply system is operating as a continuous supply. To generate 'volume-per-day' indicators following this approach, the annual volume of losses is not divided by 365 days, but by an equivalent number of pressurised days during the year. Alternatively, this can be achieved if the daily volume of losses is divided by the number of pressurised hours during the day and then multiplied by 24 hours as shown in Equation 4.3 (Charalambous and Laspidou 2017; Frauendorfer and Liemberger 2010). After this, the NRW PIs, particularly the infrastructure leakage index, can be compared to other systems with different supply patterns. The performance of the individual system should also be monitored regularly when it has changing supply times.

$$NRW_{w.s.p.} = \frac{NRW}{T_{avg}} \times 24 \qquad (4.3)$$

where $NRW_{w.s.p.}$ (m³/year) is the normalised NRW and T_{avg} is the average supply time in the system (h/day).

In this analysis, this approach is examined and extended to cover the volume of NRW, its components, and PIs. The annual volumes of losses in Sana'a were transformed into daily losses by dividing the annual volumes of the NRW components by the equivalent number of days when the system was pressurised (operated), which was 69 and 9 days in 2009 and 2015, respectively. The NRW PIs were then calculated.

Sensitivity analysis of the average supply time (T_{avg})

The sensitivity of the average supply hours during the day for the entire system, or, alternatively, the equivalent pressurised days during the year, was analysed to investigate its influence on the total volume of NRW and its PIs, and to understand the impact of T_{avg}. The influence of T_{avg} on the volume of NRW was plotted on a curve, the equation was deduced and the step-slopes of the T_{avg}–NRW curve were analysed. Accordingly, the critical points of the curve and high-sensitivity cases were determined.

4.2.6 Normalising the NRW using regression analysis

Another potential method for normalising the NRW volume and PIs is regression analysis using the NRW–SIV relationship when reliable historical data are available. Simple regression can be used to investigate the relation between the dependent variable (NRW) and the independent variable (SIV), assuming the linearity of this relationship as clarified in Equation 4.4.

$$NRW_{(SIV)} = \beta 0 + \beta 1 \, SIV + \varepsilon i \qquad (4.4)$$

where $\beta 0$ is y–axis intercept, $\beta 1$ is the slope of a straight line (expectedly positive slope) and b is the y-axis intercept. In this approach, a reasonable correlation (equation) between NRW and SIV is first established, and the NRW volume and PIs can then be normalised and calculated for any production level of the given system or zone. This equation can be used as a baseline assessment of the status of NRW in the monitored system. When the NRW status must be evaluated again for a specific year in the future, another equation is generated for that specific year, and the normalised volumes of NRW and PIs can be calculated for the same SIV to compare and assess any progression or regression in the NRW.

Extracting the actual NRW trend

The NRW–SIV regression equations were generated for 2011–2015 and the NRW levels were normalised at a certain SIV, that is, the average of 2011–2015. The actual trend of NRW was plotted and analysed to track its progress in the Sana'a water supply system during this period and compare it to the unnormalised NRW trend.

4.3 RESULTS AND DISCUSSION

4.3.1 Fluctuations in the NRW volume

Figure 4.2 shows the fluctuations in the NRW and SIV of the Sana'a water supply system under different production levels for the monthly and annual data. Figure 4.2-a shows that the NRW increases and decreases following increases and decreases in the SIV. Figure 4.2-b shows that the volume of NRW has decreased over time due to the decrease in the volume of SIV as a consequence of dwindling water resources in the Sana'a Basin. Figure 4.2-c confirms the same behaviour using the monthly data of NRW and SIV for a small district metered area (DMA) within the Sana'a water network. There are anomalies at several points in Figure 4.2-a and 4.2-c, and some data are inaccurate for some months, especially for the DMA in Figure 4.2-c at the beginning of the current conflict in Yemen, when the production and customer metering recording were unstable.

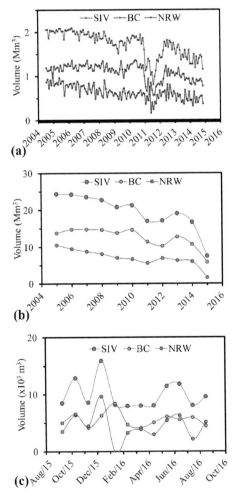

Figure 4.2. Fluctuation in the NRW volume according to changes in SIV: (a) monthly basis for the full-scale system; (b) annual basis for the full-scale system; (c) monthly basis for DMA-1 in Hadda Zone, Sana'a.

As shown in Figure 4.2, the volume of NRW varies with SIV, as it is higher or lower according to the SIV. This can also be concluded from the basic NRW equation. Obviously, the NRW volume and PIs all are affected by the volume of SIV, making NRW monitoring more difficult for intermittent supplies, as NRW PIs fluctuate according to the volume of supplied water. This concern highlights the need for normalisation when reporting the NRW level for intermittent and fluctuating supplies.

4.3.2 NRW components

The NRW component estimation is necessary to generate the different NRW PIs. The NRW components were estimated using the Apparent Loss Estimation equation elaborated in Section 6.4.4 in Chapter 6. Table 4.1 shows the volume of NRW components, 95% confidence limits (θ), standard deviation (SD), and variance (SD^2) of NRW components. The proportions of NRW components were assumed to remain the same, and these proportions are used in the analysis of 2015, as shown in Table 4.1.

Table 4.1. NRW components and uncertainties.

	Volume 2009	θ	SD	SD^2	Volume 2015
	m³/year	± %	m³/year		m³/year
NRW	8,637,692	5	227,452	5×10^{10}	1,604,557
UAC	114,152	20	11,648	1×10^{08}	21,205
AL	5,686,452	18	531,632	8×10^{10}	1,100,093
RL	2,837,088	40	578,363	1×10^{11}	483,259

4.3.3 NRW PIs

Under the current conflict situation, the leaders of Yemeni water utilities compete over the limited available fuel in local markets to deliver as much water to customers as possible. The SIV has shrunk significantly from 22.3 MCM in 2009 to 7.4 MCM in 2015 due to electricity and power shortages. The length of the mains of the network and average pressure have remained the same, at 997 km and 10 m, respectively. In contrast, the number of water connections increased from 88,936 in 2009 to 94,723 in 2015. Based on that, Table 4.2 compares the NRW volume and PIs for 2015 to those of 2009. Table 4.2 shows that all NRW PIs, expressed in different units, have reduced significantly, suggesting improved NRW performance. The PIs of the NRW components suggest the same. While all AL indicators suggest better performance, all RL indicators suggest the same, excluding the infrastructure leakage index when the intermittent supply was adjusted to be continuous (24/7) using the w.s.p. approach. This result suggests that volume-based indicators do not indicate the actual performance of the NRW status itself, but only their volumes, which vary according to changes in the SIV change. The actual NRW performance can only be reflected in the NRW PIs when the w.s.p. adjustment is extended, to cover all NRW, AL and RL indicators.

Table 4.2. NRW PIs of Sana'a for 2009 and 2015.

NRW Component	PI	2009	2015	Δ %
NRW	m³/year	8,637,692	1,604,557	−81%
NRW	%	39%	22%	−44%
NRW	m³/c/year	97	17	−83%
RL	m³/year	2,837,088	527,024	−81%
RL	L/c/d	87	15	−83%
RL	L/c/d/m pressure	9	2	−83%
ILI	-	9	2	−82%
ILI	w.s.p.	48	62	+29%
AL	m³/year	5,686,452	1,056,328	−81%
AL	m³/c/year	64	11	−83%
ALI	-	8	4	−57%

4.3.4 Normalised NRW using w.s.p. adjustment

Table 4.3 shows the normalised NRW PIs in Sana'a obtained after recalculating them using the w.s.p adjustment factor, as described in the methodology section. From Table 4.3, it can be concluded that the volumes and PIs of NRW and its components all increased significantly, indicating that the NRW status is worse, excluding the NRW as a percentage of SIV and AL index, which still indicate improvements in the NRW and AL.

The NRW % (of SIV) does not change from the figures in Table 4.2 if the SIV and billed consumption are adjusted to the same factor. However, if it is assumed that the billed consumption is not increasing, the water supply is sufficient and customers are saturated (which is not the case in Sana'a), the SIV will be adjusted and the NRW % of SIV will increase from 89% to 98% for 2009 and 2015, respectively. This further suggests worse performance, which is in line with the other PIs considered in this approach.

Similarly, for the AL index, when normalising the volumes of ALs and billed consumption to the same adjustment factor, the apparent loss index decreased from 9 to 4 in 2009 and 2015, respectively. However, if only the volume of the AL is adjusted and the billed consumption (BC) is not adjusted, the AL index also increased from 47 to 151 from 2009 to 2015, which corresponds with the other set of indicators.

Interestingly, there are great differences between the NRW PIs in Tables 4.2 and 4.3. This suggests that the NRW PIs are sensitive to the adjustment factor used in the normalisation approach. Figure 4.3 shows the sensitivity of the NRW volume normalised through the w.s.p. approach to T_{avg}, which is used as an adjustment factor. It was found that when T_{avg} decreases, NRW increases. For the power function of the curve, NRW approaches $+ \infty$ when T_{avg} approaches zero. The derivatives cannot determine the critical points of this curve. However, taking the volume of NRW at $T_{avg} = 24$ h/day as a benchmark and analysing the curve from right to left show that the volume of NRW will be doubled at $T_{avg} = 12$ h/day. Similarly, NRW will increase by 200%, 500% and 2300% when T_{avg} is 8, 4 and 1 h/day, respectively. The lower the T_{avg}, the more sensitive the normalised NRW volume. For the case of Sana'a, where T_{avg} was 4.4 and 0.6 h/day for 2009 and 2015, respectively, the normalised NRW volume becomes sensitive. Thus, the accuracy of the analysis of all NRW components and PIs is significantly influenced by the adjustment of T_{avg}.

Table 4.3. Normalised NRW PIs of Sana'a using w.s.p. adjustment.

Component	PI	2009	2015	Δ (%)
T_{avg}	h/day	4.40	0.60	−86%
T_{avg}	day/year	66.92	9.13	−86%
SIV	m³/day w.s.p.	333,106	815,500	+145%
NRW	m³/day w.s.p.	129,081	175,842	+36%
NRW	m³/c/year w.s.p.	530	678	+28%
NRW	% w.s.p*	39%	22%	−44%
NRW	% w.s.p**	89%	98%	+10%
RL	L/c/d w.s.p.	477	610	+28%
RL	L/c/d/ m pres. w.s.p.	5	6	+28%
AL	m³/c/year w.s.p.	349	446	+28%
ILI	w.s.p.	48	62	+29%
ALI	w.s.p.	8	4	−57%
ALI	w.s.p.***	45	145	+219%

* adjusted SIV and BC for w.s.p.; ** adjusted SIV and unadjusted BC for w.s.p.; *** adjusted AL and unadjusted BC for w.s.p.

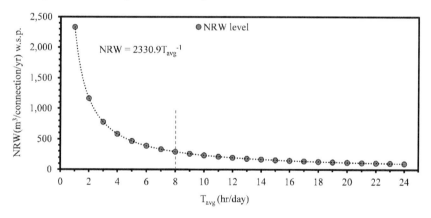

Figure 4.3. Sensitivity of the normalised NRW volume (w.s.p.) to the average supply time in Sana'a water distribution system.

However, calculating T_{avg} also has uncertainties for intermittent supply systems. In Sana'a, estimating the supply time for each distribution area within a network of 369 distribution areas is complicated, as the time of the distribution valves' closures and pumping hours of the wells and headworks must be recorded. These data are not currently available, and there would be uncertainties in their collection. Therefore, estimating the supply time for each distribution area would require significant effort and commitment. Such uncertainties significantly undermine the accuracy of normalising the NRW levels and PIs through this approach.

Additionally, w.s.p. adjustment could be extended to include AL. However, while w.s.p. adjustment is suitable for RL PIs under the same pressure and infrastructure conditions, adjusting the AL PIs leads to overestimating ALs at the point once the demand is met, and any further supplied water contributes to higher RL while the ALs remain constant. Therefore, adjusting the AL through this approach is susceptible to overestimation.

Is the NRW status progressing or regressing?

It is unclear whether the NRW situation in Sana'a has worsened, improved or remained the same from 2009 to 2015. The changes (Δ) in the NRW PIs show that, during the period of 2009–2015, the NRW level decreased by 83% of its volume and the NRW percentage decreased by 44% of its value according to the traditional approach in Table 4.2. However, when it is normalised by w.s.p adjustment, the volume of NRW increased by 28% and the value of NRW % increased by 10% if the billed consumption remained unadjusted, as described in Table 4.3. Although this 'suggested worse performance' corresponding with the poor performance of the utility during the same period due to the reduced water supply, it is still unclear whether this increase in NRW (w.s.p.) is accurate and the NRW management policies need to be revised. This could also be due to the effect of the low supply time on the normalised calculations of the NRW PIs, as highlighted in the sensitivity analysis in Figure 4.3. To verify the status of NRW, the NRW volume was normalised using another approach, through regression analysis.

4.3.5 Normalised NRW using regression analysis

The progress of the NRW status can be tracked using regression analysis for the NRW and SIV when a reasonable correlation exists. An NRW–SIV regression equation can be generated for the baseline year, and another equation should be generated for the following year, or any assessment year. The NRW volume and PIs can then be normalised at the current or a previous SIV, and the results can be compared to track the progress of the NRW's status.

Figure 4.4 shows the NRW-SIV correlations and regression equations; there was a correlation between NRW and SIV in the long term, calculated based on monthly data obtained from Sana'a's water supply system for 2005–2015 (Figure 4.4-a). The correlation is reasonable ($R^2 = 0.66$), even with the poor data obtained for some years during the analysis period. The NRW-SIV correlations were also good for annual data obtained for five years (2011–2015) for the full-scale system (Figure 4.4-b), as well as for a DMA within the network (Figure 4.4-c).

Using the NRW–SIV regression equation for 2015, as presented in Figure 4.4-h, the volume of NRW was normalised to the same production level of 2009. The results show that the normalised volume of NRW_{2015} was 9.01 MCM, while that of NRW_{2009} was 8.64 MCM. This slight difference in the level of NRW is reasonable in Sana'a and more rational than what was suggested by w.s.p. adjustment. While unnormalised NRW has exhibited a reduction in the NRW level by 81%, normalisation of the NRW level by regression analysis suggested an increase in the NRW level of 4%, and the w.s.p. approach suggested an increase of 36%.

The NRW components and PIs in 2015 can also be normalised using normalised NRW through regression analysis and compared to those of 2009. However, as the differences in the level of NRW using regression analysis are slight and the component assessment used the same proportions as those in 2009, the NRW PIs remained very close or almost the same as those of 2009, which are presented in Table 4.2.

Tracking and monitoring the NRW status, volume and PIs for an individual system are different from benchmarking and comparing the given system to other systems with different water production levels in the country or around the world. The above analysis indicates that, while regression analysis can normalise the NRW level for monitoring the NRW status of the individual system, the extended w.s.p. approach is still more useful for benchmarking and comparing different systems. However, w.s.p. adjustment suggested better performance for water supply systems with increasing T_{avg}. Linearising the curve in Figure 4.3 or developing a correction or "reduction" factor curve similar to that in Figure 4.3, but in the 4th quadrant, will be useful for conducting a more rational benchmarking of different systems with different T_{avg}.

Extracted NRW status trend

To track the behaviour of the NRW status in Sana'a over 2011–2015, NRW–SIV regression equations were generated for each year based on monthly data, as shown in Figures 4.4- d, e, f, g and h. The equivalent normalised NRW volumes for these years were then calculated at their average SIV. As the correlation factor of 2012 was not strong (Figure 4.4-e), the NRW–SIV regression equation of 2012 was not used. The behaviour of the NRW status over 2011, 2013, 2014 and 2015 is indicated by the dotted line in Figure 4.1. Accordingly, the status of NRW worsened between 2011 and 2013. This is valid for Sana'a water supply system, where the instability of the country, which began in 2011, has caused a high increase in unauthorised consumption, and sudden electricity shut-offs have caused operational problems in the network. During 2014, the situation was nearly the same, with a slight increase of the NRW due to a natural increase in leakage and limited electricity shut-offs, as the Sana'a water utility adapted to the situation. During 2015, the NRW status improved significantly, which is also valid as the Sana'a water utility valved off specific transmission pipes that the utility believed to have been illegally connected to irrigate farms along these pipes and near the well fields. In 2015, the utility also cooperated with local authorities at a district level to install isolation valves and reinstall customer water meters that had been removed in 2011. These interpretations explain well the NRW normalised curve in Figure 4.1.

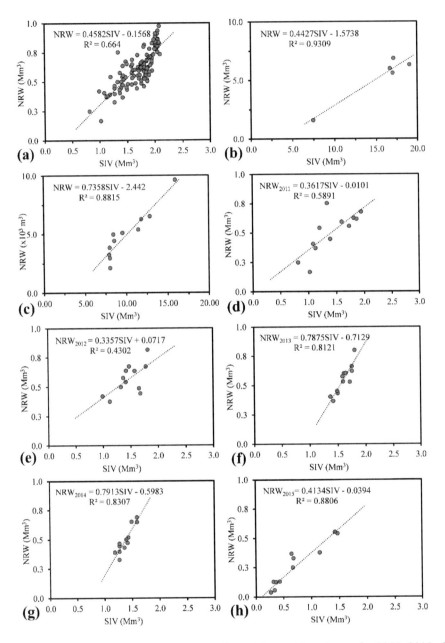

Figure 4.4. NRW–SIV regression equations; (a) monthly volumes for 2005–2015; (b) annual volumes for 2011–2015; (c) monthly volumes for a DMA in Sana'a; and (d)–(h) monthly volumes for 2011, 2012, 2013, 2014 and 2015, respectively.

4.4 CONCLUSIONS

Monitoring NRW and PIs in an intermittent water supply regime is significantly influenced by SIV. For that, a better approach is required to monitor NRW management. The influence of the varying SIV on the reported volume of NRW was investigated using monthly data of water production and billed consumption in Sana'a's water supply system for ten years. The NRW PIs were compared with and without the normalisation approach. Accordingly, the study concludes the following for intermittent water supplies:

- The volume and PIs of NRW all vary in direct proportion to the SIV. This is critical for monitoring the level and PIs of NRW for water systems with fluctuating SIV and utilities that are shifting from intermittent to continuous supplies. An increase in the NRW level does not necessarily indicate worse performance, as it could be due to an increase in the amount of supplied water. Additionally, a decrease in the NRW level does not necessarily mean better performance, as it could be due to a decrease in the supplied water. Therefore, normalisation is necessary to properly monitor and benchmark NRW PIs for intermittent supplies.

- The 'when-system-is-pressurised' adjustment, which is often used for normalising RL indicators, could be extended to normalise the volumes of NRW, AL, RL and their PIs. However, this principle leads to an overestimation of the AL, which are still difficult to monitor. This is because, when the demand is fully met, any increase in the SIV contributes to RL, while the AL remain the same. Another limitation of this approach is the sensitivity of the average supply time (T_{avg}), as its uncertainties significantly undermine the accuracy of the normalised NRW PIs, including those of the RL. In addition, this approach is likely biased towards water systems with an increasing water supply and vice versa. For water systems with a T_{avg} of less than 8 h/day, the results of this approach become more uncertain. Finally, it is not certain whether this approach indicates the actual extent of NRW progression or regression.

- For monitoring the NRW status of an individual water supply system, the NRW volume and PIs can be normalised through regression analysis. This approach reflects the actual behaviour of the NRW status and provides more rational progression and regression extents. However, this approach can only be used for monitoring the NRW for individual systems, and not for a comparison of different systems.

- Comparing and benchmarking an intermittent water supply system to other systems with high accuracy does not appear to be possible. More analysis is required to allow meaningful benchmarking using 'when-system-is-pressurised' adjustment, particularly when extending it to ALs. Until then, a correction factor curve for T_{avg} should be developed to enhance the monitoring of the NRW progression and reflect the situation of NRW for a given system among other systems with different supply patterns.

- Once a NRW monitoring tool is available, NRW management should start by network partitioning into DMAs, pressure management, active leakage detection surveys, active customer meter replacement policy and the detection of unauthorised uses. Moving towards a smart network is effective in NRW management, using smart metering, smart data acquisition and on-time acting and control.

5

MODELLING THE LEAKAGE RATE AND REDUCTION USING MINIMUM NIGHT FLOW ANALYSIS

Significant portion of the water supplied to people doesn't reach the customers but leaks out of the distribution network, causing water wastage, revenue loss and contamination risks. This chapter analyses the leakage rate, leakage components and leakage reduction potential. A minimum night flow (MNF) analysis was carried out in a district metered area (DMA) in an intermittent supply system in Zarqa, Jordan. Leakage was modelled and leakage reduction policies were analysed. Results showed that "the minimum flow" occurs in the night or in day time depending on the water levels in customer tanks, implying that a reliable one-day MNF analysis cannot be carried out in an intermittent supply system. The potential water savings of the different leakage reduction measures (pressure management; leakage detection; response time minimization) are separately analysed in the existing models in the literature, leading to overestimate the total leakage reduction potential significantly. In fact, these measures are influencing each other. Pressure reduction lowers the failure frequencies and limits the potential of leakage detection surveys, as leaks will be harder to hear and detect. Investigating the inter-dependency of these measures is therefore essential for reasonable leakage reduction modelling and planning.

This chapter has been published as: AL-Washali, T. M., Sharma, S. K., Kennedy, M. D., AL-Nozaily, F., and Mansour, H. "Modelling the Leakage Rate and Reduction Using Minimum Night Flow Analysis in an Intermittent Supply System." Water, 11(1), 48, 2019.

5.1 INTRODUCTION

All water distribution networks leak, but in different extents. In principle, leakage occurs in deteriorating networks more than in newly constructed ones, unless active leakage management and asset replacement policy are in place. As the pipes in the distribution networks reach the end of their service life, which is already the case in many countries (Gong et al. 2016), they become more vulnerable to leaks and breaks. Reducing leakage is crucial to save water, energy and revenues of water utilities, and to sustain water access to the society and the economic activities (AL-Washali et al. 2016; Dighade et al. 2014; Meseguer et al. 2014). Designing the leakage control strategy requires a baseline assessment and continuous monitoring of leakage level in the networks at full-scale as well as at a zonal-scale, or District Metered Area (DMA) (AWWA 2016; Fanner 2004; Farley and Trow 2003; Morrison et al. 2007; Thornton et al. 2008). For leakage assessment in the entire network, the top-down water balance is a common practice, where apparent losses - customer meter inaccuracies, data handling errors, and unauthorised consumption- are estimated first and then the level of leakage can be estimated from the total volume of water loss (Lambert and Hirner 2000). The top-down water balance is usually carried out on an annual basis. It gives no indications regarding the seasonal or daily leakage and lacks objectiveness in estimating the unauthorised consumption. Minimum Night Flow (MNF) analysis is the most common method for leakage assessment at the scale of the DMA. The MNF is the lowest inflow in the DMA over 24 hours of the day, depends on the consumption pattern, but reportedly occurs between 02:00 and 04:00 AM when most of the customers are probably inactive and the flow at this time is predominantly leakage (Farley and Trow 2003; Liemberger and Farley 2004; Puust et al. 2010). Several applications of MNF analysis in continuous supply systems can be found in the literature (Eugine 2017; Farah and Shahrour 2017; Latchoomun et al. 2015). Accuracy of the flow measurements and other technical considerations for MNF application are presented in Chapter 2 (Alkasseh et al. 2013; Fantozzi and Lambert 2010; Hamilton and McKenzie 2014; Werner et al. 2011). Although pressure measurement is important for leakage modelling, monitoring, and control (Alonso et al. 2000; Jowitt and Xu 1990; Lambert 2001; Thornton and Lambert 2005; Van Zyl et al. 2017; Wu et al. 2009), the volume of leakage can still be estimated using the flow and consumption data without the use of pressure gauges (Mazzolani et al. 2016; Mazzolani et al. 2017). Leakage increases with time unless controlled effectively. Successive assessments of the leakage in the network enable estimating the natural rate of rise of leakage, which is an important factor influencing the frequency of the leakage detection surveys and the pipe replacement policy (Lambert and Fantozzi 2005; Lambert and Lalonde 2005). The major portion of the leakage is avoidable, and a certain portion is unavoidable, even in the new and well-constructed network (Lambert et al. 1999; Lambert et al. 2014). However, application of MNF analysis in an intermittent supply is difficult. This is because even if a part of the network is supplied continuously for a short period (for analysis), and as long as the tanks in the DMA are not completely full, the water keeps flowing into the ground and elevated tanks in the network even if the customers are inactive during night hours. The MNF can, therefore, occur at any time other than between 2:00 and 4:00 AM. This chapter aims at analysing the minimum night (or day) flow in an intermittent supply system in Zarqa water distribution network, Jordan, where customer tanks in the DMA have to be filled, otherwise, one-day hourly flow analysis (Amoatey et al. 2018) cannot yield a satisfactory leakage estimate. The chapter also discusses the effect of upscaling the results of MNF analysis in a temporarily established DMA to the full-scale system. It also estimates the leakage components, and analyses the sensitivity of the rate of rise of leakage (RR) and infrastructure condition factor (ICF) in estimating the feasibility of the leakage reduction measures. Accurate estimation of the leakage volume, RR, and ICF triggers more reasonable leakage assessment and modelling in intermittent supplies and contributes to effective leakage reduction and control in water distribution networks.

5.2 MATERIALS AND METHODS

5.2.1 Description of the case study system

Zarqa water network serves 160,000 customers (as in 2017) which accounts for approximately one million consumer based on an average of 6.3 people served per customer. About 57% of the water comes from some 99 groundwater wells while the remaining 43% is piped long-distance from the Disi aquifer. The length of the network mains (pipes > 100 mm) is 2,447 km according to the GIS records. These represent only 30% of the network, and the remaining 70% are service connections. Various pipe materials are used, including: polyethylene, galvanised iron, ductile iron, cast iron, and steel. The network is almost fully pumped with average pressures from 10 to 60 m, except for small sections of gravity or combined supply. The water is supplied to customers through interlinked distribution areas located within five administrative zones: Rusaifah, Al-Azraq, Beerian, AL-Hashimia, and Dhulail. The water supply in Zarqa is intermittent with an average of 36 hr per week, usually during 2 days in the week. The volume of non-revenue water (NRW) in Zarqa changes every year following the fluctuations of the production level of Zarqa water utility (AL-Washali et al. 2019a), but the unnormalised average NRW volume (over the last 10 years) is 29.2 ±7.1 million m³ per year, which accounts for 57% ±4% of the supplied water.

5.2.2 DMA establishment

There are several methods for partitioning the network into DMAs (Deuerlein 2008; Di Nardo et al. 2013a; Di Nardo et al. 2013b; Galdiero et al. 2015; Galdiero et al. 2016;

Herrera et al. 2010; Kesavan and Chandrashekar 1972; Morrison et al. 2007), based on various criteria including topology, connectivity, reachability, redundancy, and vulnerability of the network. Network graph methods are common (Deuerlein 2008; Galdiero et al. 2016; Kesavan and Chandrashekar 1972; Perelman and Ostfeld 2011) and recently the design support method is suggested and applied in Monterusciello network, Italy (Di Nardo et al. 2013a). Integrating the DMAs establishment with pressure management is vital (Alonso et al. 2000; Creaco and Pezzinga 2014; De Paola et al. 2014). This research however does not design the DMAs in Zarqa network but focuses on analysing the leakage volume in a pre-set DMA under intermittent supply and its sensitivity and impacts for prioritising the leakage reductions for the entire network. To carry out the MNF analysis and estimate the volume of the leakage, a temporarily-established DMA in AL-Hashimia zone was updated (Figure 5.1), installing a separation valve, a mechanical flow meter, data loggers, and a manhole and bypass at the inlet of the DMA. The studied DMA has 1,028 customers connected to the network through 978 service connections. The mains length in the DMA is 18 km, and the total length of submains and service connection is 8.9 km. The population of the DMA was 10,426 capita in July, 2007 with an annual population growth rate in Jordan at 2.2%. Two previous attempts were made in 2002 and 2007 to carry out the MNF analysis, but were not successful as the water could not be supplied to the area for more than two days, because of a strict distribution schedule and water shortage. The water flow curves of these attempts showed unstable reading for the MNF and thus could not be used to estimate the leakage in the studied area. In the current study, the DMA was continuously supplied for 5 days starting from 2nd January 2016 at 08:00 AM till 7th January 2016 at 08:00 AM, to ensure that the studied area is fully saturated for at least one day, and the MNF readings can potentially repeat themselves.

5.2.3 Instruments and measurements

To measure the flow and pressure in the DMA, multilog data loggers (Type RDL662LFQ61-SMS made by Radcom Technologies, Dallas, TX, USA) with a memory of 48,720 readings were used. Although the loggers can be programmed to record the measurements every second, they were programmed to record the measurements every 15 min, to handle reasonable data for several days, and to conserve the batteries of the loggers, till the DMA is saturated. A new mechanical flow meter (Type: WP-Dynamic 100 made by Sensus, Hannover, Germany) with starting and maximum flows of 0.25 and 300 m^3/hr and ±2% accuracy was installed at the inlet of the DMA and connected to a data logger in a manhole that was constructed to protect the equipment. Four additional data loggers were installed to record the pressure data at four selected points attached to customer properties, to represent the pressure at different elevations within the DMA (Figure 5.2). Thus, 2,928 measurements were recorded over 5 days, 488 records for the flow measurements at 15 min. time interval, and other 488 pressure records at 15 min time interval for each of the other 5 pressure loggers.

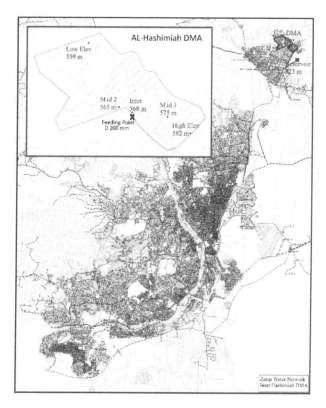

Figure 5.1. Map of Zarqa water network showing Al-Hashimiah DMA and positions of the data loggers (inset).

Figure 5.2. Data loggers used for pressure measurements.

81

5.2.4 Leakage modelling

Leakage estimation in the DMA was made using the approach presented in Chapter 2. The leakage rate at the time of the MNF was found using Equation 2.3, where the probable legitimate night consumption (LNC) is deduced from the MNF (Farley and Trow 2003; Thornton et al. 2008). Fantozzi and Lambert (2012) suggested a standard terminology for the LNC components and reviewed methods for its estimation and measurement. Automatic meter reading can be utilised to estimate the LNC accurately, if already established in the network, which was not the case in Zarqa. Hence, the LNC was estimated using the recommended assumption that 6% of the population in the DMA are active during the MNF time and that the water is used for a toilet flush (Fantozzi and Lambert 2012; Hamilton and McKenzie 2014; Thornton et al. 2008), and is in the order of 5 litres per flush. Other recommendations e.g., in UK and Germany (Fantozzi and Lambert 2012), are not applicable in Zarqa because of differences in number of people served per connection, capacity of the toilet flushes, water availability and storage, and behaviour of water consumption. The 6% assumption was validated using earlier field measurements of night consumptions in different networks in Jordan. Sensitivity analysis was further conducted to analyse the sensitivity of the estimated leakage volume to this assumption.

The leakage rate at the time of the MNF cannot be generalised for all hours of the day, because of the pressure—leakage relationship (the higher the pressure, the higher is the leakage) and variation of the pressure throughout the day. For this reason, the MNF leakage should be modelled according to the pressure—leakage relationship. In principle, a leak from an orifice in a rigid pipe can be calculated using the Torricelli equation (Equation 2.4) with a leakage exponent (N_1) of 0.5. However, the Torricelli equation fails to model the leakage in plastic pipes (Lambert 2001; May 1994) and a modified equation is proposed (Equation 2.5) (Van Zyl et al. 2017). Practitioners in the field use the Fixed and Variable Area Discharges (FAVAD) principle (Equation 2.6), which demonstrates the fact that N_1 varies from 0.36 to 2.95 and can be assumed at 1 for networks that have a mix of plastic and rigid pipes (Lambert 2001; Laucelli and Meniconi 2015; Schwaller et al. 2015; Van Zyl et al. 2017).

Using FAVAD concept in this study, leakage can be modelled at any hour during the day, assuming a fixed value for the exponent (N_1= 1) as the network is of mixed rigid and plastic pipes (Lambert 2001; McKenzie et al. 2003; Morrison et al. 2007). However, estimating the relationship between the leakage exponent N_1 and the fluctuating pressure in the DMA during the day is increasingly being considered (Di Nardo et al. 2015; Lambert et al. 2017a; Van Zyl and Cassa 2014). The zonal night test is used to determine the variable N_1 which is influenced by a changing pressure in the DMA. This is only possible when the LNC is minimal and the MNF in the DMA is almost equal to the leakage rate, which cannot be the case in this analysis. Similarly, the daily leakage rate was calculated using the night day factor (F_{ND}) as shown in Equation 2.7 and Equation 2.8 in Chapter 2 (Lambert et al. 2017a; Morrison et al. 2007), but with a correction factor of 0.97 (Lambert 2018), considering the effect of changing daily pressures in the DMA on N_I value.

5.2.5 Feasibility of leakage reductions

In principle, the economic level of leakage (ELL) can be reached when the cost to further reduce the leakage exceeds the expected benefits (Ashton and Hope 2001; Kanakoudis et al. 2012). This is because the greater the level of resources employed, the lower the additional marginal benefit which results (Pearson and Trow 2005).

From a practical perspective, the leakage consists of numerous events whose volume is a function of the run-time and flow rates for different types of the leakage (AL-Washali et al. 2016; AWWA 2016; Lambert 1994). According to the Burst and Background Estimates (BABE), the annual volume of the leakage is dominated by the run-time of the leak than the leak's flow rate. The bursts and background leakage of the entire network of Zarqa water network was modelled using a spreadsheet model (Real Loss Component Analysis a Tool for Economic Water Loss Control), developed by Water Research Foundation (USA) (Sturm et al. 2014), which uses Equation 2.10 to calculate the unavoidable volume of the leakage(Lambert et al. 2014).

The input parameters for the model were: the unavoidable leakage; the natural rate of rise of leakage (RR) set at a moderate level (3 m^3/Km mains/day/year); the variable cost of water at \$0.24/m^3; and the cost of leakage detection survey in Jordan at US\$ 100/Km (Aboelnga et al. 2018). Using these parameters, the potentially recoverable leakage is computed by the model and the optimal frequency of the proactive leakage detection surveys is estimated using Equations 3.6 and 3.7 (Lambert and Fantozzi 2005).

The potential water saving was calculated in the model for three main polices. The saving from the active leakage control was computed based on the frequency of the leakage detection surveys. The saving from minimising the response and repair time of the failures in the network was computed based on the reduction in the run times of the failures. The saving from the pressure reduction was estimated using the FAVAD principle. Eventually, the monetary value of the leakage reductions through the different polices was calculated using the variable cost of water in the network.

Furthermore, the model sensitivity to the two assumed factors, infrastructure condition and rise rate of leakage, was carried out to assess their impact on the overall economic analysis of the leakage reductions. The infrastructure condition factor (ICF) is a correction factor that ranges from 1 to 3 depending on the condition of the network infrastructure compared to that of the typical cases upon which the BABE model was developed and is reflected in the parameters of Equation 2.10. (Fanner and Thornton 2005). The ICF factor was analysed at two recommended levels (1.5 and 2.5). The Rise Rate of leakage (RR), is a factor that indicates the normal rate at which leaks would increase in the network if there is no leakage control policy. The RR impact on the output of the model was analysed using low, moderate and high levels (Fanner and Lambert 2009). Finally, the model was run using the estimated leakage by carrying out MNF analysis in one DMA in the network and generalising it for the entire network. The model was also run using the average leakage from two methods, the MNF analysis and the top-down water balance, which should be more representative for the entire network. The model outputs were then compared.

5.3 RESULTS AND DISCUSSION

Figure 5.3 shows the pressure and flow measurements in the DMA during the experiment. Figure 5.3a shows the logged pressures at three points: the inlet point, a mid-elevation point and a high-elevation point. These ranged from 6.0 to 60.0 m with a mean of 32.4 m and a standard deviation of 14 m.

Figure 5.3b shows the measured pressures at two other points that lie between the same range in Figure 5.3a, but with a pressure drop at mid 2 point on Monday, 4 January 2016 from 9:00 AM till 18:00 PM. This pressure drop coincided with flow drop from 43 to 5 m³/hr on the same day (Figure 5.3c) due to a change in a valve situation, but shortly recovered in the records of the next 15 minutes. The flow and pressure drops are highlighted in the right-side of Figure 5.3. The experiment, however, succeeded to reach the saturation level on the following 3 days which are used for the MNF analysis: Tuesday, Wednesday and Thursday 5, 6, and 7 January 2016 respectively. The saturation status was reached on the 3rd day, after 63 hours (2.63 days) of continuous supply. By that time the water entering the DMA satisfied the demand and all ground and elevated tanks were full, and the records started to closely repeat themselves for 3 consecutive days, enabling analysis of the leakage rate in the DMA. Figure 5.3c shows the typical demand—pressure relationship in the DMA where the pressure (mean) drops down when there is high demand and rises when there is less demand, during night or morning time. Figure 5.4 shows the leakage-pressure modelling based on the flow and pressure measurements. Firstly, the LNC was deduced from the MNF rate, and then the leakage rate of this specific time was calculated. As the MNF time lasts for 1.5 hour, LNC was divided by the same value, to get the hourly LNC which was used to calculate the leakage rate at the MNF time. Secondly, the hourly leakage volume during the day was modelled as shown in Figure 5.4 using the FAVAD concept. Finally, the daily leakage volume was calculated based on the F_{ND} (Equation 2.7 and 2.8).

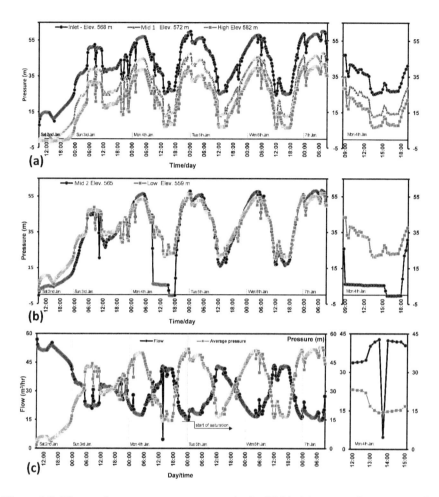

*Figure 5.3. Flow and pressure measurements in the DMA: (**a**) range of pressure in the DMA; (**b**) pressure measurement in further two points in the DMA with pressure collapse in one point; (**c**) flow and average pressure relationship in the DMA.*

The leakage volume in the DMA was 882 m^3 for the period of the experiment, which is 24.1% of the supplied water. This is based on the assumption that 6% of the population are active during the MNF time. The validity of this assumption was verified using field investigation of LNC in several networks in Jordan. Furthermore, the sensitivity of the estimated leakage to this assumption is also analysed in the supplementary data of this chapter. Nevertheless, Table 5.1 shows that if the LNC is altered or divided by two hours (instead of 1.5 hour), the leakage increases to 927 m^3 (25.4% of the supplied water). This indicates the sensitivity of the results to the LNC. In this analysis, the values used for LNC were 2.51 m^3/hr for 753 people using 5 L flush toilets during a period of 1.5 hour.

85

The values of the MNF used to calculate the daily leakage were almost similar on 5 and 7 January, and the calculated night day factor was 14.2 hr/day. However, if the average MNF of the last three days is used, the leakage would be 25.1% of the supplied water, indicating less sensitivity of the results than LNC. Additionally, Figure 5.5 also shows that MNF time occurred at 12.15 AM, 4:45 AM, and 7:15 AM for the last three days of the experiment respectively. Similar findings were reported in the literature (Alkasseh et al. 2013; Lambert et al. 2017b). This means that MNF can also occur even when only some of the customers are active (e.g. for Fajr prayer at 5:00 AM in January), provided that no water is pumped from the ground tanks to the elevated tanks during the MNF time.

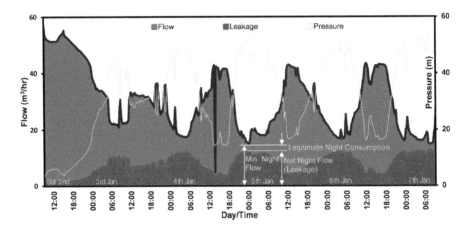

Figure 5.4. Leakage modelling in AL-Hashimia DMA, Zarqa.

Table 5.1. Sensitivities of the parameters of leakage volume estimation.

Date	MNF	MNF time	LNC	LNC duration	NNL	NDF	Daily leakage	Leakage volume	Leakage level
	m³/hr	AM	m³/hr	hr	m³/hr	hr/day	m³/day	m³	% SIV
5 Jan.	15.0	12:15	1.88	2.0	13.1	14.2	185.8	932.7	25.5%
			2.51	1.5	12.5	14.2	176.9	888.1	24.3%
6 Jan.	16.4	4:45	1.88	2.0	14.5	14.2	205.0	1029.4	28.2%
			2.51	1.5	13.9	14.2	196.2	984.8	26.9%
7 Jan.	14.8	7:15	1.88	2.0	13.0	14.2	183.5	921.3	25.2%
			2.51	1.5	12.3	14.2	174.6	876.7	24.0%
Avg.	15.4	-	1.88	2.0	13.5	14.2	191.4	961.2	26.3%
			2.51	1.5	12.9	14.2	182.6	916.6	25.1%
Avg. 5&7 Jan.	14.9	-	1.88	2.0	13.0	14.2	184.6	927.0	25.4%
			2.51	1.5	12.4	14.2	175.8	882.4	24.1%

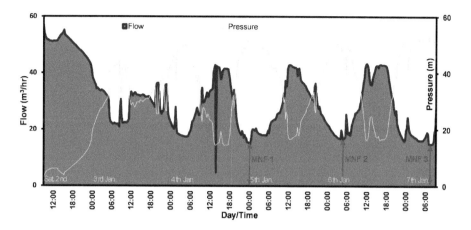

Figure 5.5. Time of MNF occurrence in the DMA.

Generalising the leakage level of Al-Hashimia DMA to the entire network depends on how representative the DMA is and the extent of uncertainty regarding similarities and differences of network assets and operating conditions. The higher the number of DMAs modelled, the more representative will be the estimated level of leakage for the whole network, but probably fully representative only if the entire network is divided into DMAs and MNF is carried out for all the DMAs. Even though, to have an annual estimate of the leakage level, MNF analysis should be regularly carried out throughout the year for all the DMAs, as leakage could vary with the time. This is not technically, operationally, and economically possible in the current situation in Zarqa network. For this reason, the leakage level of the DMA was assumed to represent the entire Zarqa network, and further investigation was carried out.

Based on the MNF analysis, the leakage level of the network was estimated at 16.1 million m³/year. Further analysis for the leakage components was carried out using the BABE approach and the break and failure records for each pipe diameter in the network. The response time to repair the reported or detected leaks was computed by the Maintenance Management System (DCMMS) and it averaged 2 days in 2014. About 261 leaks detected through leakage detection surveys were estimated for each pipe diameter along with estimated awareness and repair times (183 and 2 days, respectively). Accordingly, the leakage from reported and unreported failures was estimated at 2.4 million m³/year. The background leakage was estimated at 1.8 million m³/year, based on the concept of the unavoidable annual real losses (Equation 2.12). The differences between the estimated leakage volume by the MNF analysis and the sum of these two volumes is considered as hidden losses and it amounted to 11.8 million m³/year. It is not known how much of the hidden losses are recoverable and how much are unavoidable. Obviously, the component analysis of leakage (BABE) involves a small part of the leakage in this case. Although the ICF was assumed at high level in the model (ICF= 2.5), the component analysis of the leakage model analysed only 26.3% of the leakage in Zarqa network while the remaining 73.6% is thus considered as hidden losses. This is probably due to the empirical assumptions of flow and characteristics of bursts and unavoidable leakage in the BABE model that are not totally applicable for all cases. With an ICF value of 5, the hidden losses remained more than 60% of the leakage. This result emphasises the need for a future research on the use of the BABE model in the intermittent supply context.

Feasibility of leakage reductions

Based on the component analysis of leakage, savings resulting from leakage reduction measures were analysed using the model "Component Analysis: a Tool for Economic Water Loss Control". Different scenarios for leakage reductions are possible in Zarqa. Figure 5.6 shows the potential water and monetary savings through different measures. Reducing the average response time of repairing reported and detected leaks from 2 days to 3 hrs is a target of Zarqa utility. This would save 1.9 million m^3/year with a monetary value of US$ 0.4 million, using the variable production cost in Zarqa. Adopting Active Leakage Control (ALC) policy through regular leak detection surveys that covers the network every 10.5 months could save 10.7 million m^3/year (US$ 2.5 million). Reducing the average pressure of the entire network from 33 to 23 m; e.g., through separating the elevated parts of the network; would save 4.9 million m^3/year (US$ 1.2 million). However, we note that the economic model analyses each of these measures independently while in reality it is likely that these measures are dependent on each other. For instance, pressure reduction limits the failure frequencies and lowers the potential of ALC as leaks will be harder to detect. For this reason, aggregating the possible savings from the three measures is likely to overestimate the potential impact of the leakage reductions; e.g., the savings will be higher than the volume and value of the leakage. For this reason, future modelling of economic benefits of leakage reduction measures should take into account the inter-dependency of these measures.

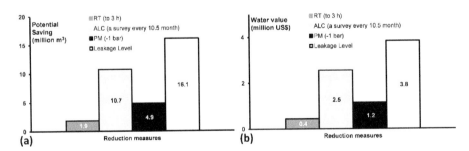

Figure 5.6. Potential water saving of different leakage reduction measures. (a) volume of water saved (b) value of water saved.

To gain an insight into the critical factors in the aforementioned economic model, the sensitivity of two factors were analysed. The RR was manipulated in the model for low, moderate and high values (Fanner and Lambert 2009) and then the resulting savings were reported. This step was conducted for two ICF levels, (1.5 and 2.5). Figure 5.7a shows the impact of changing the RR on the water saving (left axis) and the monetary value (right axis) when adopting ALC through leakage detection surveys. At low level of RR (e.g. 1 m³/km mains/day/year), 66% of the network should be surveyed every year. The potential saving will be about 11.2 million m³/year (US$ 2.7 million) with a survey annual cost at US$ 0.15 million based on a survey cost of US$ 100/km. At high level of RR (e.g. 5 m³/km mains/day/year), 147% of the network should be surveyed every year. The potential saving will then be 10.4 million m³/year (US$ 2.5 million) with a survey annual cost at US$ 0.33 million. For moderate RR (3 m³/km mains/day/year), 114% of the network should be surveyed every year with a saving of 10.7 million m³/year (US$ 2.5 million). This analysis demonstrates the high sensitivity of RR value when modelling the economic frequency of leakage detection surveys, which complicates the task in intermittent supplies where RR is difficult to estimate accurately. For this reason a moderate level of RR was used in Zarqa modelling and Figure 5.7a shows the model results for different RR values. Figure 5.7b shows the same parameters of Figure 5.7a but for ICF at 1.5, which is not very sensitive in the output of the economic model, and the figures are close to those in Figure 5.7a. Clearly, more research work is needed to improve the reliability of the economic analysis of leakage through fixing the RR and the component analysis of leakage in intermittent supplies.

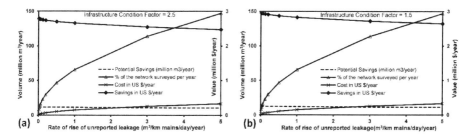

Figure 5.7. Sensitivity of RR at ICF of 2.5 and 1.5 (graphs a and b)

Finally, to get a more reasonable feasibility analysis of leakage reductions for the entire network, the overall volume of leakage was estimated through the top-down water balance and integrated with the MNF analysis (Thornton et al. 2008). Apparent losses were estimated as elaborated in Appendix A1 and then the leakage volume for the entire network was calculated and compared with the estimated leakage from the MNF analysis. Figure 5.8 shows the model results when the volume of leakage in the model is changed to 26.9 million m³/year, which is the average of the leakage volumes as estimated through

MNF analysis and through the top-down water balance. While MNF analysis is reasonably accurate at the DMA-scale, upscaling its results for the entire network is uncertain and sensitive. Leakage estimation from one DMA is not likely to be sufficiently representative for an entire network. The overall annual volume of leakage in a network should be verified through several methods before being used for leakage reduction modelling. In all cases, estimating the benefits of the ALC seems to be overestimated when using Equation 3.6. Further investigation is required to confirm this assessment, and also to clarify the interdependency between potential savings of ALC and pressure management.

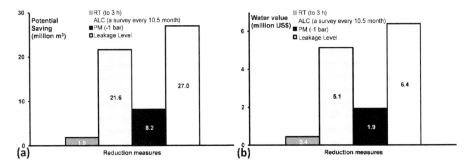

Figure 5.8. Potential savings achieved due to leakage reduction, as estimated by top-down water balance and MNF methods (a) volume of water saved (b) value of water saved.

5.4 CONCLUSIONS

5.4.1 Minimum night flow analysis in intermittent supplies

- One -day minimum night flow analysis (MNF) cannot be used to estimate the leakage rate in networks of intermittent supply because water keeps flowing during night time to fill customers' tanks in the network. Therefore, the examined zone (DMA) should be supplied continuously for several days till the zone is saturated, the customer ground or elevated tanks in the network are completely full, and the minimum flow readings start to closely repeat themselves.

- In networks of intermittent supply, the "MNF" could occur at night or day time, even if some of the customers are active, provided that the ground tanks are full and customers do not pump water from the ground to the elevated tanks. In Zarqa, the saturation of the DMA started after 63 hours of continuous supply and the "MNF" was taking place between 12:00 AM and 7:00 AM. Facing this challenge requires more careful estimation of the legitimate night consumption, which was found to be an important and sensitive parameter in leakage estimation and modelling.

- Generalising the leakage rate at the time of the MNF and using it for the entire day overestimates the daily leakage, because the pressures during the day are usually lower. For this reason, the night day factor in Zarqa is a reduction factor (<24 hr/day), being 14 hr/day.

- While MNF analysis is reasonably accurate at a DMA-scale, upscaling the results for the entire network is uncertain and sensitive. One or several DMAs cannot make-up for the diversity of the operating conditions in the network in terms of pressure, flow rates, pipes' lengths, and number of connections. Therefore, estimation of the leakage of an entire network should be verified through several methods before using it for full-system leakage reduction modelling.

5.4.2 Leakage reduction modelling in intermittent supply networks

- The leakage component analysis model (BABE) analyses only a small part of the leakage (26% in the Zarqa case) while the remaining 74% is considered as hidden losses where the recoverable and unavoidable leakage are not known. Increasing the Infrastructure Condition Factor was not sufficiently influential in the case study. Therefore, the (BABE) model may require further research to adapt it for the intermittent supply context.

- Analysing the potential water savings under different leakage reduction policies, taken one by one, is currently possible. However, this approach is likely to overestimate the potential saving significantly, due to the inter-dependency of the different policies. Sometimes it may even show potential savings to be more than the volume and cost of the leakage. In all cases, the benefits of the frequent leakage detection survey seem to be over-estimated, and further investigation is required to clarify and confirm this issue.

- The inter-dependency between pressure management and active leakage control should also be investigated. Pressure reduction limits the failure frequencies and lowers the potential of leakage detection as leaks become harder to detect. Therefore, future leakage reduction modelling should closely examine the influence of various leakage reduction policies on each other (e.g., pressure management on ALC).

5.5 SUPPLEMENTARY DATA

5.5.1 Sensitivity analysis of leakage level in a DMA in Zarqa for different assumptions of LNC

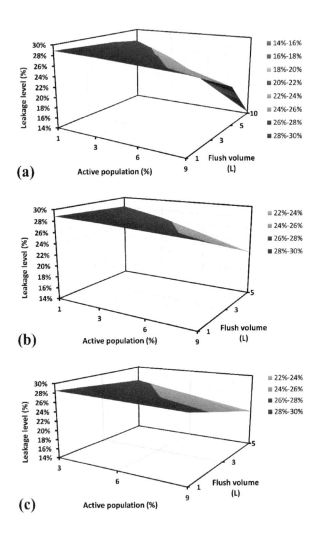

Figure 5.9. Sensitivity of the leakage level in a DMA in Zarqa under different assumptions of LNC (a) all possible assumptions, (b) removing the assumption of 10 L flushing capacity as the common flushing capacity in Zarqa is less than 10 L, (c) at three possible assumptions of % of active population and three levels of flushing capacity.

Figure 5.9 indicates that the leakage level ranges from 21% to 30% but is mostly about 27% of the supplied water. The estimated leakage is 24.1% of the supplied water with LNC that is equivalent to 0.2 L/person/hr. An earlier field survey conducted by Zarqa utility found LNC to be around 1.2 L/subscriber/hr, which indicates the same level of LNC. The uncertainty of the LNC component to the volume of the leakage ranges from ±10% to ±20% of its value. To use this analysis in Chapter 7, the assigned uncertainty of different uncertainty components of this analysis is assumed at ±30%.

6

ALTERNATIVE METHOD FOR NON-REVENUE WATER COMPONENT ASSESSMENT

This chapter presents a practical method for estimating the volume of apparent losses in water supply systems, particularly in developing country context, so that real losses can be calculated from the volume of Non-Revenue Water (NRW). The method employs the fact that almost all apparent losses reach wastewater treatment plants (WWTP). The volume of apparent losses can be estimated through establishing a water and wastewater balance using routine measurements of WWTP inflows and routine data of billed consumptions. The water and wastewater balance was applied in the water supply system of Sana'a, Yemen and NRW components were estimated. Sensitivity and uncertainty analyses show that all sources of the errors are small apart from the level of accuracy of measuring the WWTP inflow, and this accuracy can be practically improved. Other parameters and variables are relatively insensitive. The chapter discusses the advantages and limitations of the suggested alternative method compared to the other NRW component assessment methods.

This chapter has been published as: AL-Washali, T. M., Sharma, S. K., and Kennedy, M. D. "Alternative Method for Nonrevenue Water Component Assessment." Journal of Water Resources Planning and Management, 144(5), 04018017, 2018.

6.1 INTRODUCTION

Since 2000, NRW assessment has progressed significantly due to large efforts made by the International Water Association (IWA) and other organizations to promote new concepts and methods for NRW management (Mutikanga 2012; Vermersch and Rizzo 2008). The main methods of NRW assessment at a component level are the bottom-up Minimum Night Flow (MNF) analysis and the Top-Down Water Balance. Burst And Background Estimates (BABE), which is also referred to as the component analysis, is considered to be more of a real loss component analysis than a NRW component assessment tool. A detailed review of these methods is presented in Chapter 2. However, there are certain contexts that limit the application of each of these methods. MNF is difficult to apply where water supply is intermittent or insufficient, and where isolation of District Metered Areas (DMAs) has not been implemented. MNF analysis requires capital investment where the DMAs are not established, as well as sophisticated equipment and technical capacity. It is difficult to apply to representative DMAs that can provide general assessment for the level of real losses for the entire network, on a regular basis. Application of the BABE model on its own is not recommended because of a significant level of uncertainty in much of the data used in the analysis (Thornton et al. 2008). The assumptions embedded in generating BABE factors are based on data for cases that have certain contexts, technologies and policies of active leakage control. The Top-Down Water Balance methodology requires estimating the unauthorised consumption – which is challenging in cases where unauthorised consumption is significant. There is no systematic methodology to estimate this component, especially for illegal use by unregistered domestic and agricultural users. It should be noted that assuming the unauthorised consumption at a certain level is not useful for its monitoring. Therefore, developing a methodology that is capable of assessing NRW components in such contexts is worthwhile especially where apparent losses are significant or dominant, and MNF analysis cannot be conducted regularly for representative DMAs.

This chapter introduces a method to assess NRW at component level and to break down the volume of NRW into apparent losses, real losses and unbilled authorised consumption. The method determines the volume of apparent losses through establishing water and wastewater mass balance, and then, real losses can be calculated. The chapter presents the underlying principle, developed equations, advantages and limitations of this "alternative" method, and it demonstrates an application of the method in the case study of Sana'a, the capital city of Yemen. The chapter contains three main parts: (i) an overview of the alternative method; (ii) the development of the method; and (iii) an application example of the method in Sana'a, Yemen.

6.2 OVERVIEW OF THE METHOD

The underlying principle of this method is that almost all apparent losses enter sewer networks. Tracking water from its production site to consumers confirms that a certain fraction of the actual consumption enters the sewer network. This also includes water used through illegal connections and bypasses. Meter inaccuracies and data handling errors that are reflected on billing data do not appear in the wastewater flow. Therefore, studying the inflows of the wastewater treatment plant (WWTP) could indicate the quantity of the apparent losses. Figure 6.1 shows a theoretical water and wastewater balance that can be used to estimate the apparent losses. This figure also shows that a certain part of real losses do not enter the sewer network. Other portions of leaks that enter or infiltrate to the sewers is considered with the volume of infiltration/inflow to the sewers, and in certain cases can be a negligible quantity. Once water is used (some outdoors and the rest indoors), wastewater is generated. The wastewater flow is generated from authorised water customers, unauthorised water users, customers with bypasses and customers with malfunctioning or inaccurate meters. Therefore, apparent losses can be estimated by utilizing the data of WWTP inflows or the outflow of a certain DMA.

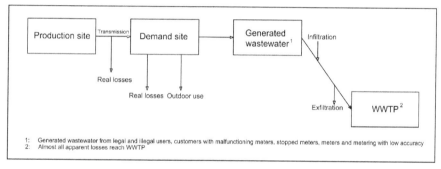

Figure 6.1. Conceptual illustration of the water and wastewater balance

This balance is applicable where the wastewater collection and treatment system exists and is centralised. It considers the WWTP dry weather inflows, so the rain inflows have no effect in the balance. Where grey water is reused, it is either reused indoors and enters the sewers, or reused outdoors so it contributes to the outdoor water use. The method can also be applied in systems with a small coverage of wastewater service. However, it is more accurate where the majority of water customers are covered with wastewater services, so the balance is representative for the entire system.

6.2.1 The Apparent Loss Estimation (ALE) equation

The volume of apparent losses can be monitored regularly and calculated through the Apparent Loss Estimation (ALE) equation (Equation 6.1). In the ALE equation, the volume of apparent losses is a result of only two variables: the inflow to WWTP and the amount of billed water.

$$Q_{AL} = (A + 1)Q_{ww} - (B - C + 1)Q_{bw} \qquad (6.1)$$

where Q_{AL} is the apparent losses (m³/year) if the assessment period is one year, Q_{ww} is the inflow of the WWTP (m³/year) and Q_{bw} is the billed water (m³/year). A, B and C are case-specific factors; A is the exfiltration-infiltration factor (3% – 10%), B is the unbilled authorised consumption factor (0.5% - 1.5%) and C is the outdoor water use factor (4% - 40%). There are three steps to assess the volume of apparent losses using this equation: (i) setting the equation factors and fixing them; (ii) obtaining the data of the billed water and WWTP inflows; and (iii) calculating the volume of apparent losses. The apparent losses can be monitored regularly through ALE equation based on measurements of billed water and inflows of WWTP. The real losses can then be calculated by subtracting the volume of apparent losses from the total volume of water loss.

However, there are two points to be considered for application of the ALE equation. Firstly, the rain inflows that enter the sewers and eventually increase the WWTP inflows should be discounted. The WWTP inflows on rainy days throughout the assessment period should be substituted by the average dry weather flows of the remaining dry days (m³/day). Secondly, there can be cases in which wastewater service coverage is less or more than water supply service coverage. In such cases, the obtained data of "billed water" and "WWTP inflows" should be adjusted for customers with both water and wastewater services. Adjusting the billed water can be achieved through Equation 6.2. Similar equations are applied to adjust WWTP inflows to customers with water and wastewater services.

$$Q_{bw-adjusted} = \frac{Q_{bw} \times N_{w\&ww}}{N_w} \qquad (6.2)$$

where Q_{bw} is the volume of billed water, $Q_{bw-adjusted}$ is the volume of adjusted billed water to customers with both services (water and wastewater), N_w is the number of customers with water service and $N_{w\&ww}$ is the number of customers who have water and wastewater services. The effect of the large customers in Equation 6.2 should be reasonable or the exhaustive data should be used. The volume of apparent losses obtained from the ALE equation should then be used to recalculate the volumes of apparent losses and real losses for all water customers. This can be achieved by dividing the volume of apparent losses for customers with water and wastewater services to the number of these customers. The result should then be multiplied by the total number of water customers to find out the total volume of apparent losses for the entire water supply system.

6.2.2 Setting the equation factors

The factors A, B and C in ALE equation are case-specific factors that should be set first and then monitoring the level of apparent losses is possible. These factors represent the rate of exfiltration and infiltration to the sewers, the volume of unbilled authorised consumption and the rate of outdoor water use respectively. For initial assessment of the apparent losses, factor A can be assumed at 5%, factor B at 0.7% and factor C at 15% (0.05, 0.007 and 0.15 respectively) because the parameters are not highly sensitive to the output of ALE equation. However, the exact values and sensitivities of these factors should be checked, as there are differences from a system to another.

The exfiltration factor A ranges from 3% to 10% of the wastewater base flow. The infiltration to the sewers should be estimated and considered in factor A and thus its value is reduced, or can be neglected where it is believed to be negligible. For initial assessment, exfiltration can be assumed at 5% – or different values of factor A can be assumed and then optimised based on analysing the outputs of the ALE equation. The range of factor A depends on different influences that affect the rate of exfiltration out of the sewers such as: sewer age, depth, diameters, permeability of the surrounding soil, suspended solids, organic load and quality and types of sewer pipes. For example, the level of self-sealing due to cohesive solids in sewage should be higher in sewers with high level of suspended solids and organic matter, and thus exfiltration rate should be lower. Factor A can also be verified through Equation 6.3 where exfiltration and infiltration to the sewers can be calculated based on billing data and "measured" per capita consumption that must not include any types of water losses.

$$Q_{ex} - Q_{inf} = N_p \times Q_{cap}(1 - Q_{out}\% \div 100) + Q_{ind} - Q_{ww} \qquad (6.3)$$

where Q_{ex} is the volume of exfiltration, Q_{inf} is the volume of infiltration/ inflow, N_p is the number of population covered with wastewater service, Q_{cap} is per capita water consumption, $Q_{out}\%$ is the outdoor use percentage out of water consumption, Q_{ind} is the industrial and commercial wastewater discharge and Q_{ww} is the WWTP inflow.

The factor B is a factor representing the volume of the unbilled authorised consumption. It is usually a small amount that has no significant reflection on the ALE equation. It can be assumed from 0.5% of billed water as suggested by Lambert et al. (2014) to 1.25% of system input volume (SIV) as recommended by AWWA (2009). The factor B can also be estimated based on available data in a water utility and then converted to a percentage of billed water. The factor C represents the volume of outdoor water use and should be estimated through Equation 6.4 based on monthly billing data, as will be elaborated in Section 6.3.1.

$$Q_{out\%} = \frac{Q_{bw} - 12 \times q_{bw.min.month}}{Q_{bw}} \times 100 \qquad (6.4)$$

where $Q_{out\%}$ is the outdoor water use percentage, Q_{bw} is the annual volume of billed water and $q_{bw.min.month}$ is the volume of billed water in the minimum month of the year.

6.2.3 Theoretical example

If the SIV of a certain supply system is 60,000 m³/day, the billed water is 35,000 m³/day, and the WWTP inflow is 38,000 m³/day, the volume of apparent losses can be estimated through ALE equation by first setting the factors of the equation, and then calculating the volume of apparent losses. Assuming or estimating the exfiltration rate, factor A at 7%; the outdoor use, factor C at 10%; and the unbilled authorised consumption, factor B at 1%, then measuring and monitoring the volume of apparent losses in this supply system can be regularly conducted through Equation 6.5.

$$Q_{AL} = 1.07 \, Q_{ww} - 0.91 Q_{bw} \qquad (6.5)$$

In this example, the volume of NRW is 25,000 m³/day (41.7% of SIV). The apparent losses are then 8,810 m³/day (14.7% of SIV or 35.2 % of NRW), the unbilled authorised consumption is 350 m³/day (0.6% of SIV or 1.4% of NRW), and real losses are 15,840 m³/day (26.4% of SIV or 63.4% of NRW). To give an overview of how the ranges of factors A, B and C affect the volume of apparent losses in this example, Figure 6.2a shows the range of apparent losses as a percentage of SIV corresponding to the ranges of factors A, B and C. The volume of apparent losses ranges from 12%-17% of SIV as factor A ranges from 3% - 10%; it ranges from 14% - 13% of SIV as factor B ranges from 0.5% - 1.5%; and it ranges from 11% - 16% of SIV as factor C ranges from 5% - 15%. Consequently, considering the minimum and maximum ranges of factors A, B and C, the minimum volume of apparent losses is 11% of SIV, the maximum volume of apparent losses is 17% and the average is 14%. Therefore, the sensitivity of apparent losses due to ranges of A, B and C factors is reasonable. Figure 6.2b shows an example of sensitivities of these factors in the case study of Sana'a.

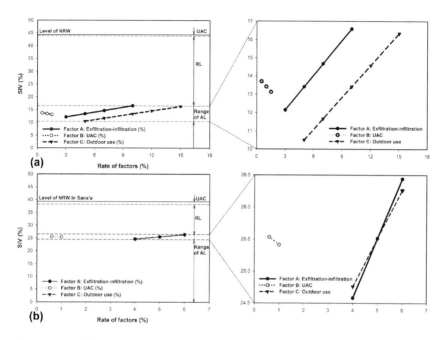

Figure 6.2. (a) Range of apparent losses as % of SIV for different rates of A, B, C factors in a theoretical example; (b) Sensitivity of volume of apparent losses to the error margins of factors A, B and C in Sana'a water supply system

6.3 DEVELOPMENT OF **ALE** EQUATION

In light of the water and wastewater balance shown in Figure 6.1, and referring to Equations 6.6, 6.7, 6.8 and 6.9, the volume of apparent losses can be calculated from Equation 6.10.

$$NRW = Q_{AL} + Q_{RL} + Q_{uac} \qquad (6.6)$$

$$NRW = SIV - Q_{bw} \qquad (6.7)$$

$$SIV = Q_{RL} + Q_{out} + Q_{ww} + Q_{ex} - Q_{in} \qquad (6.8)$$

Substituting Equation 6.8 in Equation 6.7

$$NRW = Q_{RL} + Q_{out} + Q_{ww} + Q_{ex} - Q_{in} - Q_{bw} \qquad (6.9)$$

Substituting Equation 6.6 in Equation 6.9

$$Q_{AL} = Q_{ww} - Q_{bw} - Q_{uac} + Q_{out} + Q_{ex} - Q_{in} \qquad (6.10)$$

where, Q_{AL} is the volume of apparent losses, Q_{RL} is the volume of real losses, Q_{ww} is the inflow of the WWTP or the flow in the sewers' outlet of a specific zone, Q_{bw} is the billed water consumption, Q_{uac} is the unbilled authorised consumption, Q_{out} is the outdoor water use, Q_{ex} is the exfiltration; volume of wastewater that ex-filtrates out of the sewers and Q_{in} is the infiltration/inflow; e.g. groundwater that infiltrates into the sewer system.

6.3.1 Variables investigation

There are six variables to be determined in Equation 6.10. The WWTP inflows are usually measured flows at an hourly or daily basis at the inlet of WWTP, and the billed water and unbilled authorised consumption are routine measurement data in any water utility. In contrary, the last three variables need further investigation.

Outdoor water use

Outdoor residential use consists of water used for lawn irrigation or watering of plants, car washing, house cleaning and refilling of fountains, ponds and surface lagoons. The outdoor water use could be estimated by using variables that reflect the outdoor characteristics such as irrigable area of the premises, garden sizes and pool ownership (Arbués et al. 2003). Interestingly, the amount of water used for irrigation purposes is the dominant proportion of outdoor use (Gleick et al. 2003; Palenchar et al. 2009; Singh et al. 2009). Mitchell et al. (1999) estimated garden irrigation at 90% of the outdoor water use. If so, the climatic conditions' variance influences the outdoor irrigation and therefore, the outdoor water use (Mitchell et al. 1999). For this reason, the methods of estimating the outdoor water use are based on the conclusion that irrigating areas of the premises makes up almost all of the outdoor water use proportion.

There are different methods for estimating the outdoor water use (Mini et al. 2014). Some methods estimate outdoor water use using landscape coefficients and estimated irrigation requirements, based on landscape characteristics and reference evapotranspiration. Some methods rely on the formulation of urban water balance models that require climate data, land cover characteristics, surface retention capacities, soil storage capacity, field capacity, water use data, water storage conditions and surface aerodynamic characteristics of evapotranspiration. Other methods use total indoor water use to derive the outdoor water use estimate as a residual. These methods use water billing data or direct measurements through household logged water data and flow trace analysis. For estimating the outdoor water use in this chapter, the methods that rely on routine billing data were chosen due to applicability and data availability.

There are three methods of estimating outdoor water use that rely on billing data. Historically, outdoor water use has been estimated by subtracting the winter consumption from the metered consumption (Mayer et al. 1999). In the summer-winter method, the difference between winter consumption (October to March) and summer consumption (April to September) is approximately equal to outdoor use (Gleick et al. 2003; Romero and Dukes 2010). On the other hand, Gleick et al. (2003) used minimum month method and average month method. The minimum month method is the most popular way for calculating the outdoor water use (Palenchar et al. 2009). In the minimum month method, the lowest use month is assumed to represent indoor use and all differences between the other months and minimum month is considered to be outdoor use (Romero and Dukes 2010). The underlying assumption of this method is that indoor use remains fairly consistent across seasons. This assumption is verified by Mayer et al. (1999) that carried out a field study on 1200 study homes in 14 cities in REUWS, USA. The study used historical billing data for the whole year and logged indoor usage data for two periods in summer and winter with two weeks each. It concluded that indoor use remains consistent for 13 out of 14 cities, and there are no statistically significant differences in indoor use across seasons. In the average month method, the average of the three lowest water consumption months is computed to be equal to indoor use and outdoor use is calculated as the residual (Mini et al. 2014). Nevertheless, Skeel and Lucas (1998), as reported by Gleick et al. (2003), found that outdoor water use for single family houses accounted more than 95 percent of the observed increase in peak summer consumption and less than 5% was due to a slight increase of the indoor use in summer months. This slight increase in indoor use during summer will be counted as outdoor use, which leads to overestimating the annual volume of outdoor use by approximately 1 percent. This amount can be neglected as several studies highlighted that methods of estimating outdoor use based on billing data are likely to underestimate the volume of actual outdoor use due to the probable existence of landscape irrigation during the minimum month(s) consumption, as reported by Gleick et al. (2003), A&N Technical Services Inc et al. (2013), and Mini et al. (2014). For the purpose of estimating apparent losses through water and wastewater balance, the minimum month method is used as it is more accurate than the other two methods due to the potential use of outdoor water during winter months.

Sewers exfiltration

There are several methods for estimating the sewer exfiltration rate. These methods include: quality monitoring of groundwater, wastewater mass balance, pressure tests for single pipes in the field, tracers and lab investigation. Rutsch et al. (2008) reviewed the methods available for estimation of exfiltration and evaluated their accuracies. Ellis et al.

(2003) concluded from experimental studies that exfiltration from urban sewers can be estimated as no more than 5% - 10% of average daily dry weather flow (base wastewater flow). Ellis et al. (2008) studied the main factors affecting exfiltration rates including the formation of clogging and bio-film layers, soil type and permeability, and pressure head of exfiltration. They pointed out that magnitude of exfiltration is still unclear and a comprehensive solution for the assessment of sewer exfiltration does not seem to be yet at hand. They suggested a unifying framework to facilitate focused model building. Rutsch (2006) approached the general exfiltration rate at 3% and listed the exfiltration rate reported in several studies ranging from 1% - 13% of dry weather flow. For the purpose of this study, based on the literature review, the exfiltration rate can be assumed at 3% - 10% of dry weather flow depending on the factors that influence the exfiltration rate.

Sewers infiltration/inflow

Inflow and infiltration (I/I) both increase the flows in urban sewerage systems. Inflow is water entering the sewer system directly through sources such as manhole covers, surface connections, sub-surface flow through the unsaturated zone, and land drainage connection. For separate sewer systems, inflow includes the mis-connected direct runoff to the sewer system. It is usually associated with extreme wet weather conditions and therefore represents a fast response component of sewer flows. On the other hand, infiltration is water entering the sewer system from groundwater or below groundwater level through openings such as displaced joint pipes, cracks and breaks in the fabric of sewers, manholes and chambers. It is a slow response process resulting in increased flows mainly due to elevated groundwater entering the drainage system (Dublin Drainage Consultancy 2005; Ellis and Bertrand-Krajewski 2010). For the purpose of this study, the amount of rain inflow can be discounted through counting the dry days WWTP inflows and use their average for the rainy days throughout the assessment period. The groundwater infiltration to the sewers can be neglected where the saturated zone or groundwater table is far below the sewers. When the sewers are within the saturated zone, the infiltration rate can be estimated in a similar way to the MNF analysis. Groundwater infiltration can be estimated by analysing the data of dry weather WWTP inflows during the period from 02:00 – 04:00 a.m. in which domestic wastewater connections are likely to be inactive. It is then assumed that 10% of the wastewater flow at the minimum period of night time is legitimate and the rest is caused by infiltration (Dublin Drainage Consultancy 2005).

ALE equation

From the previous section, it can be concluded that the exfiltration rate can be a function of the wastewater base flow (exfiltration = f($Q_{wastewater}$) = $A \, Q_{wastewater}$), and the infiltration should be considered as a reduction on the exfiltration factor as shown in Equation 6.3, or neglected if it is insignificant. The unbilled authorised consumption can also be considered as a function of billed water (unbilled authorised consumption = f($Q_{billedwater}$) = $B\,Q_{billedwater}$). The outdoor water use can be considered as a function of billed water (outdoor use = f($Q_{billedwater}$) = $C\,Q_{billedwater}$) as explained above. Therefore, Equation 6.10 can be re-arranged as in the ALE equation presented in Equation 6.1.

6.4 APPLYING THE METHOD IN SANA'A, YEMEN

Sana'a city is the capital and largest city in the Middle East country of Yemen, which is a water-scarce country. The supply system in Sana'a was constructed in the 1970s, and the network has been extended with rudimentary ad-hoc expansions to meet rapid population growth of the city. Sana'a water supply system covers 45% of the population, and the rest are supplied by the private sector mainly via water tankers. The utility serves 94,723 customers which account for 1.5 million consumers. The only source of water is a deep aquifer where water is extracted from 114 wells with depths from 600 – 1000 meters below the ground surface. The length of the network mains is 977 km. The mains are constructed of ductile iron, unplasticised PVC, and asbestos-cement pipes with diameters varying from 150 – 800 mm. The submains and service connections are constructed from galvanised iron and high-density polyethylene. The system in Sana'a is a combined system with both pumped and gravity supply. Around 30% of the distribution network is pumped directly from headworks with a pressure from 40 to 60 meters at the headworks, 20% is pumped directly from the wells, and around 50% of the distribution network is supplied by gravity from elevated tanks and reservoirs within the network. The network is divided geographically into six administrative zones, and these zones are subdivided into 369 distribution areas that are interlinked and multi-fed. The supply pattern in Sana'a is intermittent and insufficient. A customer receives the water once a week for an average rate of 4.4 hours per day. Water is supplied randomly to the distribution zones with no equity among customers, and thus consumers compete for the water in the network. If the water provided via the public network is insufficient, customers have to buy supplementary water from private water tankers at a higher price. The level of NRW in Sana'a is high, and the average level for 10 years is 35% without considering the production meter inaccuracies and the potential overbilling practices.

6.4.1 NRW assessment in Sana'a

Assessment of the NRW components in Sana'a is challenging. Yet, there are no DMAs in Sana'a as there is inadequate technical knowledge about the network components and lack of technical capacity within the utility. However, even if there were DMAs, MNF analysis could not be applied because the water supply in Sana'a is intermittent and insufficient; therefore, customer tanks in a DMA cannot be saturated. On the other hand, the assumptions of the top-down water balance can neither be applied nor monitored in

Sana'a where apparent losses are dominant. Assuming the unauthorised consumption in Sana'a at 0.1% of supplied water as recommended or used in European utilities (Lambert and Taylor 2010; Vermersch et al. 2016), at 0.25% of supplied water as recommended by AWWA (2009), or at 10% of billed water, or 10% of NRW as suggested by Mutikanga et al. (2011a) and Seago et al. (2004), is not objective nor does it allow regular monitoring of unauthorised consumption in Sana'a. Estimating the individual components of unauthorised consumption is also a tedious task that requires time and resources (AWWA 2009). There is no suggested methodology to estimate the components of the unauthorised consumption, especially the illegal use by unregistered users for domestic and irrigation uses. For these reasons, the water and wastewater balance has been developed for Sana'a water utility. To apply this method, data of Sana'a water supply system have been obtained. Since 2011, Yemen has been facing an unsettled situation which affects water supply and water services in Sana'a, as there is a shortage of fuel and electricity. For this reason, the only complete set of available data of Sana'a is for the year 2009 which was set to be the assessment period.

Volume of NRW

The volume of NRW is the difference between SIV and the billed water for the assessment period. The assessment period was set for one year (2009) as it is long enough to include seasonal variations and it reduces the effects of lag time in customer meter readings (Austin Water Utility 2009). The SIV was adjusted for production meter inaccuracies, which were assumed at 7% ±2% under-registration based on the estimation of the production unit in Sana'a water utility. After produced water is adjusted to production meters inaccuracies, and the billed water is adjusted to time lag between production and billing, the volume of NRW is calculated. The metered and unmetered unbilled authorised consumptions were inventoried and estimated for each administrative zone in Sana'a water utility.

6.4.2 Estimating the volume of apparent losses

Estimating the volume of apparent losses in Sana'a was achieved through: (i) determining the A, B and C factors in ALE equation; (ii) obtaining the billed water and inflows of WWTP; and (iii) estimating the volume of supplementary water supply in Sana'a as the supplied water by Sana'a water utility is not sufficient for many areas in the city, and some customers buy supplementary water from private water tankers to cover their water demand. Part of this supplementary supply is transferred through the sewers to the WWTP and should be considered in the water and wastewater balance.

The exfiltration factor A ranges from 3% to 10% of the wastewater dry weather flow. This was verified in Sana'a using Equation 6.3 with per capita consumption at 61.4 L/day (MWE 2007). The resulting rate was 7.1% which is overestimated and represents the maximum level of exfiltration as the per capita consumption used is not measured and thus includes water losses. Therefore, the exfiltration rate in Sana'a was assumed at 5% ±1% of wastewater base flow. The groundwater infiltration to sewers in Sana'a was neglected because water table is more than 600 m deep; other potential inflows/infiltration to the sewers have been considered in Equation 6.3. Source of errors and sensitivity analysis are conducted to evaluate the influence of assuming this parameter at this level. The factor B was calculated in Sana'a at the level of 0.8% of billed water based on the records of Sana'a water utility.

The factor C in Sana'a was assumed at 5% ±1% of billed water because the minimum month method cannot be applied using the billing data of Sana'a water utility that does not represent the actual variance of seasonal consumption. This is because the supply is insufficient and there is a supplementary supply via water tankers. To justify this assumption, outdoor use percentage in Dhamar city in Yemen was calculated at 11% based on consumption data using the minimum month method. Figure 6.3 shows the monthly variance of outdoor use for Dhamar during 2009. The minimum month (February & June) is assumed to represent indoor use, and the differences between February and other months is considered to represent the outdoor use. However, the outdoor water use in Sana'a should be lower due to the limited areas of yards and irrigation land in the premises of the city. Mitchell et al. (1999) highlighted the link between housing density and outdoor water use due to the size of the premises' gardens; outdoor use generally increases with spacious blocks and decreases with high-density units in the city. Therefore, the percentage of outdoor water use calculated for Dhamar city was reduced by 50% as suggested by water management experts in Water and Environment Centre in Sana'a University for two reasons: (i) house yards and irrigation areas of the premises in Dhamar city are larger and more prevalent; and (ii) water supply is continuous (24/7) and sufficient, and thus outdoor use is higher. The impact of this assumption was investigated by analysing the sensitivity and influence of the value of factor C on the overall accuracy of ALE equation.

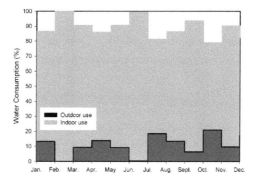

Figure 6.3 Monthly outdoor water use % during the year 2009 for Dhamar city, Yemen.

The volume of the supplementary supply in Sana'a was estimated based on the production records for the private supply wells within the service area according to the 2010 database of Sana'a wells in the National Water Resources Authority.

Estimating and monitoring apparent losses in Sana'a

After factors *A*, *B* and *C* are set and the supplementary supply in the city is considered in the water and wastewater balance, estimating and monitoring of apparent losses in Sana'a can be conducted regularly using Equation 6.11.

$$Q_{AL} = 1.05Q_{ww} - 0.96Q_{bw} - 0.95Q_{ss} \qquad (6.11)$$

where Q_{ss} is the volume of supplementary supply in the service area. The billed consumption as well as inflow measurements of Sana'a WWTP at its inlet were obtained and adjusted for only water and wastewater customers using Equation 6.2. The calculated volume of apparent losses per customer was then used to generate the total volume of apparent losses for all customers in Sana'a water supply system. The volume of real losses was then calculated, based on the above.

6.4.3 Results and discussion

Volume of NRW components

Non-revenue water in Sana'a water distribution system makes up 38.75% of SIV and stands for 8,637,692 m³/year. The total volume of the unbilled authorised consumption is 114,152 m³/year, which makes up 0.5% of SIV, 0.8% of billed water and 1% of NRW. Meanwhile, data was obtained in order to estimate apparent losses in Sana'a through Equation 6.11: wastewater inflows as measured at the inlet of Sana'a WWTP, at a total of 19,361,500 m³/year; billed consumption for "water and wastewater" customers, at a total of 12,760,667 m³/year; and the supplementary water supply during the assessment period, at a total of 2,931,898 m³/year. Therefore, the volume of apparent losses in Sana'a water supply system is estimated using Equation 6.11 at 5,314,944 m³/year for those customers who have both water and wastewater services, 68 m³/year per customer, and 5,686,452 m³/year for all water customers. The volume of real losses is then calculated at 2,837,088 m³/year. The breakdown of NRW into components shows that the shares of apparent losses, real losses and unbilled authorised consumption are 66%, 33% and 1% of NRW respectively. These figures are close to the expectations of Sana'a water utility (apparent losses around 60% and real losses around 40%). The IWA standard water balance for Sana'a water supply system for the year is shown in Figure 6.4.

Supplied Water 22.29 Mm³ ±0.45 100.0%	Authorized Consumption 13.77 ±0.03 61.8%	Billed Authorized Consumption 13.65 ±0.01 61.3%	Billed Metered Consumption 13.51 ±0 60.6%		Revenue Water 13.65 ±0.015 61.3%
			Billed Unmetered Consumption	0.15 ±0.015 0.7% *	
		Unbilled Authorized Consumption 0.11 ±0.02 00.5%	Unbilled Metered Consumption	0.04 ±0.002 0.2% *	Non-Revenue Water 8.64 ±0.45 38.7%
			Unbilled Unmetered Consumption	0.07 ±0.02 0.3% *	
	Water Losses 8.52 ±0.45 38.2%		Apparent Losses 5.69 ±1.04 25.5%		
			Real Losses 2.84 ±1.13 12.7%		

Figure 6.4 IWA standard water balance in million m³/year and % of SIV for Sana'a water supply system. Starred cells are not scaled; exaggerated height.

To compare the results of this method to another traditional method, apparent losses have been estimated differently. The customer meter inaccuracies were estimated at -4.15%, under-registration, based on a random sample of 22 meters collected from the field and tested under critical flow rates; the data handling errors were estimated at -5.7% of the billed water based on estimations of the water utility and available data of a campaign conducted for this purpose; and the unauthorised consumption was assumed at 0.25% of the supplied water as recommended by AWWA (2009). Accordingly, Table 6.1 shows a comparison of the volumes of NRW components and NRW performance indicators based on these two methods. The levels of apparent losses and real losses differ significantly for each method and were reflected on the NRW performance indicators. Estimating the apparent losses using the unauthorised use assumption of AWWA (2009) underestimates the level of apparent losses in Sana'a and gave unacceptable estimate for Sana'a water utility.

The revenues of Sana'a water utility in 2009 was 1,672 million Yemeni Rail (YR) (USD $7.78 million as of February 2016). With total billed water at 13,652,622 m³, the average revenue is 122.5 YR ($0.57)/m³. The production cost in 2009 was 69.4 Y.R. ($0.32)/m³ (SWSLC 2010). Based on these data the cost of NRW was calculated at 904 million YR ($4.2 million) of which 77% is apparent losses, 22% is real losses and 1% is unbilled authorised consumption. Figure 6.5 shows the volume and value of NRW components.

Table 6.1. Selected NRW performance indicators of Sana'a water supply system for different NRW component estimation methods.

Component	Performance Indicator	W&WWB This study	UC=0.25% of SIV (AWWA 2009)
NRW	(m³/yr)	8,637,692	8,637,692
	(%)	39%	39%
	(m³/connection/yr)	97	97
	(m³/connection/yr) w.s.p.	530	530
Apparent Losses (AL)	(m³/yr)	5,686,452	1,426,406
	(% of NRW)	66%	17%
	(m³/connection/yr)	64	16
	(m³/connection/yr) w.s.p.	349	87
	ALI	8	2
Real Losses (RL)	(m³/yr)	2,837,088	7,097,134
	(% of NRW)	33%	82%
	(L/connection/d)	87	219
	(L/connection/d) w.s.p.	477	1,193
	(L/connection/d/m pressure)	9	22
	ILI	9	29
	ILI w.s.p.	48	159

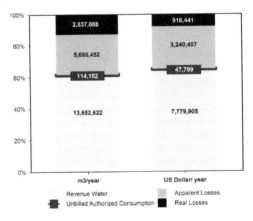

Figure 6.5 Volume and value of NRW in Sana'a, Yemen (US Dollar = 214.97 Yemeni Rial as of February 2016).

6.4.4 Error and uncertainty analysis

There are uncertainties in estimating apparent losses in Sana'a through ALE equation. The production meter inaccuracies were assumed to be -7% ±2% of SIV, based on the estimation of the production unit. As 99% of the billed water in Sana'a is metered, it is assumed that there is no error margin in the volume of metered billed water, because customer meter inaccuracies and errors in data handling of customer meter readings are considered in the component of apparent losses. Therefore, the error margin of volume of NRW is the same as that of SIV, which is ±5% of the volume of NRW. The random and systematic errors associated with estimating the amount of unbilled authorised consumption is ±20% of its volume.

The source of errors associated with estimating apparent losses is substantial. There are three variables in Equation 6.11. The accuracy of WWTP inflow measurements can be within ±5% of its measured volume as ultrasonic flow meter is used (MultySonic 8000). The analytical human errors of the volume of supplementary supply were assumed to be ±10% of its volume. The error margins of factors A, B and C were assumed at ±20% of their values. Therefore, the aggregated and propagated uncertainty of the estimated apparent losses in Sana'a can be calculated through error propagation theory by Equation 6.12 at ±18% of the volume of apparent losses.

$$U_{AL} = \frac{\sqrt{\Delta Q_{ww}^2 + \Delta Q_{bw}^2 + \Delta Q_{uac}^2 + \Delta Q_{out}^2 + \Delta Q_{ex-inf}^2 + \Delta Q_{ss}^2}}{Q_{AL}} \qquad (6.12)$$

where U_{AL} is the uncertainty percentage of apparent losses and Δ is the volume of uncertainty of the variables of the water and wastewater balance. The aggregated margin of error of real loss is then 40% of its volume. The best precision achieved for calculating volume of real losses is ±20% (Lambert 2009) and ±10% (Lambert et al. 2014) in fully metered systems; while Sana'a is a complicated case.

Analysing the uncertainty of apparent losses, its breakdown showed that 60% of this error margin is a consequence of the inaccuracy of the WWTP measurements, 18% is for supplementary supply uncertainty and 22% is due to the uncertainties associated with assumptions of factors A, B and C. If there were no supplementary supply in the city, the error margin would be 75% for WWTP measurement inaccuracies and 25% for uncertainties of factors A, B and C. This analysis shows that the first priority for improving the accuracy of the results is not related to factors of the ALE equation or to the supplementary supply estimation. Rather, the priority is to improve the accuracy of the WWTP inflow measurements. Table 6.2 shows 95% confidence limits; α, the standard deviation; σ and the variance σ^2 of variables of the water and wastewater balance with two levels of the WWTP inflow inaccuracy; ±5% and ±2%. The uncertainty of apparent and real losses can be reduced by 48% and 38% respectively when the accuracy of the WWTP inflow measurements is reduced from ±5% to ±2%; which can be achieved in Sana'a by increasing the crosswise measurements of the ultrasonic meter from 2 to 4 paths. This precision level should be acceptable in Sana'a for several reasons: other methodologies to estimate apparent losses are difficult to apply in Sana'a; initial water balance usually has a high level of uncertainty, then data improvement for the water

111

balance is a process; and the precision of WWTP inflows can be improved or calibrated and its level on the uncertainty can be minimised.

Table 6.2. Statistical parameters of the water and wastewater balance.

Component	Volume m³/yr	α ±%	σ m³/yr	σ²	α ±%	σ m³/yr	σ²
Q_{ww}	19,361,500	5%*	968,075	9.37E+11	2%	387,230	4.00E-04
Q_{bw}	13,652,622	0%	14,622	2.14E+08	0%	14,622	2.14E+08
Q_{out}	784,628	20%*	156,926	2.50E+10	20%	156,926	2.50E+10
Q_{ex-inf}	968,075	20%*	193,615	3.70E+10	20%	193,615	3.70E+10
Q_{ss}	2,931,898	10%*	293,190	8.60E+10	10%	293,190	8.60E+10
NRW	8,637,692	5%	446,046	1.99E+11	5%	445,806	1.99E+11
Q_{uac}	114,152	20%*	22,830	5.21E+08	20%	22,830	5.21E+08
AL	5,686,452	18%	1,042,000	1.00E+12	10%	546,389	2.99E+11
RL	2,837,088	40%	1,133,685	1.00E+12	25%	705,553	4.98E+11

* Assumed uncertainties

Figure 6.2b was generated to emphasise the low sensitivity of the volume of apparent losses in Sana'a due to error margins of assumptions of factors *A*, *B* and *C*. When one of the factors *A*, *B* or *C* is increased by 20% of its value, the corresponding increase in the volume of apparent losses is 3.6%, 0.4% and 3.0% of its volume respectively. The lines of factor *C* and factor *A* in Figure 6.2b coincide, the influence of the factors' error margin is not sensitive as the range of apparent losses in Figure 6.2b is not wide.

6.5 COMPARISON WITH OTHER METHODS

The water and wastewater balance method is not meant to be a "one-size fits-all" method, nor does it replace the conventional methods of water loss component assessment. Rather, the method is an attempt to contribute to the understanding of water loss components and report experience gained from the case of Sana'a. However, compared to MNF analysis, the proposed alternative method does not require intensive field work, nor advanced knowledge about network components, nor sophisticated metering equipment, nor high technical capacity, nor night flow and consumption estimation. The method employs routine data and therefore can be carried out regularly. The time and sample size of water and wastewater balance is large enough and more representative for the entire network throughout the assessment period than one or several DMAs within the network. Unlike MNF analysis, the water and wastewater balance method is not pressure-dependent; this is significant because average pressures are sensitive and questionable.

Compared to estimating NRW components through BABE factors, estimating apparent losses through water and wastewater balance is not dependent on real loss estimation, which has high levels of uncertainties. An example of these uncertainties is the average pressure for the entire network, which is difficult to evaluate. Unlike the BABE factors, this method does not contain assumptions that are drawn from specific cases and generalised for other systems. The quality of pipe materials and construction, the level of leakage detection technology, the operating conditions, and the assumption that water utilities carry out a policy of active leakage control are not always applicable with the same level of the studied cases that generated the BABE factors. In addition, all underlying equations of the water and wastewater balance are presented so the variables' accuracy can be evaluated, adapted and improved.

Compared to the top-down water balance, the introduced method can be used to generate more objective estimation for the volume of apparent losses. The water and wastewater balance integrates with the top-down water balance and enhances its accuracy. The method can be a solution for developing countries where the volume of apparent losses is significant and the top-down water balance low assumptions of apparent losses cannot be applied. Furthermore, the water and wastewater balance is the first method that enables water utilities to estimate the volume of unauthorised consumption objectively without the need to estimate real losses.

6.6 CONCLUSIONS

An alternative method for estimating the volume of apparent losses based on water and wastewater balance was developed and applied in Sana'a water supply system. The volume of apparent losses can be estimated through this method using the ALE equation (Equation 6.1), based on routine measurements of WWTP inflows and routine data of billed water. The NRW in Sana'a for the assessment period was 38.7% ±2% of SIV of which 25.5% ±4.7% is apparent losses, 12.7% ±4.7% is real losses and 0.5% ±0.1% is unbilled authorised consumption.

Sensitivity and uncertainty analyses showed that the accuracy of measuring WWTP inflows, which was ±5%, contributes to 60% of error margin of the volume of apparent losses. If the WWTP inflow measurements and volume of NRW are 100% accurate, then the error margin of the volume of apparent losses and real losses through this method in Sana'a would be ±1.7% of SIV, including uncertainties of supplementary supply, exfiltration, infiltration and outdoor use. All sources of errors are relatively small compared to the level of accuracy of measuring the WWTP inflows. Other parameters and variables are relatively not sensitive. All components of error margin are not inherent in the method itself, but associated with the input data.

However, some limitations of the method were reported. The high sensitivity of the accuracy of WWTP inflows remains the main limitation, yet it can be practically improved by accurate metering equipment. Calculating factor A through Equation 6.3 is susceptible to sources of errors of per capita consumption. Expectedly, the uncertainties of exfiltration, infiltration and outdoor water use estimations should be not sensitive. However, a sensitivity graph (similar to Figure 6.2) should be made and then a decision can be reached on the use of the results from this method. In addition, for developed countries where apparent losses are not significant, this method requires testing; ALE equation parameters should be modelled first, and then the ratio of wastewater to billed water per connection can indicate the volume of apparent losses on a regular basis.

7

COMPARATIVE ANALYSIS OF WATER LOSS COMPONENT ASSESSMENT METHODS

Reducing all water loss components to zero is neither technically possible nor economically viable. The water loss components should, therefore, be accurately assessed and prioritised for their reduction. This chapter investigates the four methods that break down the water losses in distribution networks into apparent and real losses. Their accuracies and uncertainties are discussed and applications to three case studies in developing countries are presented. The results show that different methods estimate the water loss components differently. Consequently, different reduction measures are planned and prioritised. Interestingly, the least accurate methods have a low level of uncertainty, but more realistic assumptions yield higher uncertainties. This suggests that the uncertainty analysis only assists in improving the outputs of each of the methods but does not demonstrate their accuracy. The cost of water loss varies depending on the assessment method used and the economic feasibility of the reduction measures is significantly influenced. The water loss components should, therefore, be assessed for the whole network using at least two methods to reasonably model and monitor the loss reduction in water distribution networks.

_____

This chapter has been published as: Al-Washali, T., Sharma, S., Lupoja, R., Al-Nozaily, F., Haidera, M., and Kennedy, M. "Assessment of Water Losses in Distribution Networks: Methods, Applications, Uncertainties, and Implications in Intermittent Supply." Resources, Conservation and Recycling, 152(1), 104515, 2020.

7.1 INTRODUCTION

Reducing all WL components to zero is neither technically possible (Lambert et al. 2014) nor economically feasible because the greater the level of the resources employed is, the lower are the additional marginal benefits (Ashton and Hope 2001; Kanakoudis et al. 2012; Pearson and Trow 2005). After the economic level of WL, any further investment does not result in cost-effective water savings, excluding the environmental costs and impacts of water abstraction (Ashton and Hope 2001; Molinos-Senante et al. 2016). To effectively and efficiently minimise the WL, the WL should be diagnosed, its components and subcomponents should be assessed, and their reduction should be prioritised (Kanakoudis and Tsitsifli 2014; Mutikanga et al. 2012; Puust et al. 2010). However, comprehensive and well-accepted methods for WL assessment were not available two decades ago (Liemberger and Farley 2004). Later, significant advancements were made due to the development of new concepts and methods for WL management (Mutikanga 2012; Vermersch and Rizzo 2008). As shown in Table 7.1, the components of WL, real losses (RL) and apparent losses (AL), can be assessed using the common top-down water audit methodology (AWWA 2016; Lambert and Hirner 2000) or, alternatively, by establishing a water and wastewater balance (AL-Washali et al. 2018). Leakage can also be estimated using Minimum Night Flow (MNF) analysis (Eugine 2017; Farah and Shahrour 2017; Farley and Trow 2003; Puust et al. 2010) or the component analysis of the leakage (AL-Washali et al. 2016; AWWA 2016; Lambert 1994). Yet, these methods use different approaches (and scales) to estimate the WL components and thus different corrective measures are prioritised (AL-Washali et al. 2019b) and different economic levels of leakage are planned, contributing to less effective WL management. A detailed review of three methods of water loss component assessment is presented in Chapter 2. This chapter, however, investigates applications, and uncertainties of WL component assessment of four methods using three case studies in developing countries: Zarqa, Jordan; Sana'a, Yemen; and Mwanza, Tanzania. Subsequently, the sensitivity of the WL component assessment for planning WL reduction interventions, particularly leakage reduction, is analysed. The results can be used to enhance the accuracy of WL component assessments and facilitate more reasonable planning and more effective WL management in water distribution networks, especially with respect to intermittent supplies and developing countries.

7.2 DESCRIPTION OF THE CASE STUDY SYSTEMS

The water loss assessment methods were applied in three case studies: Zarqa, Jordan; Sana'a, Yemen; and Mwanza, Tanzania. The Zarqa water supply system serves 160,000 customers. It is an intermittent water supply system with an average supply time of 36 h per week.

Table 7.1. Scale, approach, and limitations of water loss component assessment methods

Method	Reference	Scale	Approach	Limitations
Top-down water audit	Lambert and Hirner 2000	system-wide	- assume and estimate AL components and then calculate RL. - desk method; pressure-independent; cost-effective - cost-effective	- focus on RL not AL; generic assumptions of AL - no methodology to estimate unauthorized consumption - likely overestimates RL
Water and wastewater balance	AL-Washali et al. 2018	system-wide	- estimate AL using WWTP inflow measurements and then calculate RL - desk method; pressure-independent; cost-effective	- requires centralised sewers for all or part of the network. - needs measurements of WWTP inflows
MNF analysis	Farley and Trow 2003	District Metered Area (DMA) scale	- estimate leakage in a part of the network - both assessment and reduction process; actual measurements - pressure-dependent	- intensive field work, zoning - requires trained manpower and sophisticated equipment - estimates leakage in a part of the network during a time of the year
Component analysis of leakage (BABE)	Lambert 1994	system-wide	- analyse field data and volumes of bursts and the rates of small background leaks - the only method that breaks down RL into subcomponents, cost effective - clarifies the nature of leakage and simulates its reduction; pressure-dependent	- applicable only for utilities that have regular active leakage control (ALC) - many assumptions; underestimates RL; further calibrations are useful

The average non-revenue water (NRW) level in Zarqa is 29.2 million cubic metre (MCM), accounting for 57% of the system input volume (SIV). Further details about this case study system are conferred in Chapter 5. The Sana'a water supply system serves 94,723 customers. It is also an intermittent water supply system with an average supply time of 4.4 h per week. The average NRW level in Sana'a is 7.1 MCM, accounting for 35% of the SIV. Further details about this case study system are presented in Chapter 6. The Mwanza water supply system serves 49,284 customers (as in 2015), with 15.5 people per customer, that is, ~0.77 million users. The main source of water is raw water from Lake Victoria with an elevation difference of 74 m. The length of the network mains is 870 km. The submains are constructed from ductile and cast iron, polyvinyl chloride, high-density polyethylene, and polyethylene pipes with diameters ranging from 25 to 500 mm. The water in Mwanza is almost continuous pumped supply, with an average supply time of 22 h/d. The supply network has been divided into five zones and several separated DMAs among which few contain flow meters to measure the inflow to these areas. Based on obtained field data, the average NRW level in Mwanza for the period of 2009–2015 was 14.3 MCM, accounting for 48% of the SIV.

7.3 APPLICATION OF THE METHODS

Water balance

The top-down water balance method was applied in the three case studies. For the Zarqa water supply system, the customer meter inaccuracies were estimated by Zarqa water utility based on a lab bench test for a sample of customer meters for different float-valve flows of the tanks. The data handling errors were estimated by extensive audits of the billing data of the water utility conducted by the authors (Appendix A1). On the other hand, two assumptions were made regarding the unauthorised consumption (UC), that is, 0.25% of the SIV (AWWA 2009) and 10% of the billed water (Mutikanga et al. 2011a), which is close to other recommendations for developing countries (Seago et al. 2004; Wyatt 2010). Based on these two assumptions, two different AL volumes were estimated; accordingly, two different RL volumes were calculated from the WL volume.

Similarly, the customer meter inaccuracies for the Sana'a water supply system were estimated by the authors based on a lab bench test on 22 customer meters representing different types, ages (or registered readings), and sizes. To have an insight on the field customer meter accuracy, the sample of the meters should be tested under the field and float-valve flows (AL-Washali et al. 2020a). Measurements of the network flows were obtained from the utility, and the flows of the float-valve were experimented from its fully open status to the closure level, with the network inflow. Based on Bernoulli's principle, in the fully open status of the float-valve, the flow that passes the customer meter is the network's flow. When the float-valve starts to partially close, the flow that passes the water meter is the flows of the float-valve. The samples were collected from the field and tested under these flows representing the actual flows in the field. The meters' accuracy was estimated for different heights in the tank and different openings of the float-valve, and accordingly the weighted meter accuracy was estimated. The data handling errors were estimated using utility data based on a sample audit conducted by the Sana'a water utility. Two assumptions were made with respect to the UC, similar to Zarqa, and two AL volumes were estimated. Accordingly, two RL volumes were calculated. The same methodology was also applied for the Mwanza water utility but based on a sample of 30 customer meters collected from the field and tested for two flows programmed in the bench test equipment.

Water and wastewater balance

The water and wastewater balance method was applied in only two cases, that is, for the Sana'a and Mwanza water supply systems. For Zarqa, the WWTP inflow data were not accessible because the WWTP was operated by the private sector. The application of the water and wastewater balance method to the Sana'a water supply system is discussed in Chapter 6. Firstly, factors A, B, and C in the Apparent Loss Estimation (ALE) equation (Equation 6.1) for the Sana'a water supply system were set to 5%, 0.8% and 5%, respectively. These factors were set based on Equation 6.3, Equation 6.4, and the data obtained from Sana'a water utility. Uncertainties and sensitivities of these factors are discussed in Chapter 6. Because the Sana'a water utility provides insufficient water to its customers, the supplementary supply through water tankers (Q_{ss}) was added to the balance and the AL was then estimated using the ALE equation, as presented in Equation 7.1 (AL-Washali et al. 2018)

$$Q_{AL\ Sana'a} = 1.05Q_{ww} - 0.96Q_{bc} - 0.95Q_{ss} \qquad (7.1)$$

where Q_{AL} is the volume of apparent losses (m^3/year), Q_{bc} is the volume of billed consumption (BC) (m^3/year), and Q_{ss} is the volume of supplementary supply by water tankers (m^3/year). The AL volume calculated using Equation 7.1 only accounts for customers with both water and wastewater services. The AL rate per customer was calculated and then generalised for all water customers to obtain the total AL volume for the whole network. Subsequently, the RL volume was calculated from the total WL volume. Factor A in the ALE equation (Equation 6.1) for the Mwanza water supply system was assumed to be 7%; the low sensitivity of assuming this factor is discussed in Chapter 6 and AL-Washali et al. (2018). Factor B was estimated to be 0.63% based on utility data audits and factor C was calculated to be 9% using Equation 6.4. Accordingly, the AL in Mwanza was estimated using the ALE equation, as shown in Equation 7.2, and available data for only four months and for customers with both water and wastewater services. The AL rate per customer was calculated and generalised for all water customers. The RL volume was then calculated from the WL volume.

$$Q_{AL\ Mwanza} = 1.07Q_{ww} - 0.92Q_{bc} \qquad (7.2)$$

Minimum night flow analysis

The MNF analysis was only applied in two cases: Zarqa and Mwanza. The application of the MNF method for the Sana'a water supply system was not possible because of the failure in completely separating a DMA and because the supplied water is not sufficient to reach the saturation condition, where all ground tanks in the DMA are completely full and the DMA becomes a continuous pressurised system during the test. For the Zarqa water supply system, the authors established a temporary DMA in the AL-Hashimia zone, which contains 1,028 customers that are linked to the network via 978 service connections as elaborated in Chapter 5. For the measurement, an isolation valve, mechanical flow meter, and four pressure loggers were installed in the DMA and water was continuously supplied for five consecutive days from 08:00 am on 2 January 2015 to 8:00 am on 7 January 2016. It is believed that the DMA was saturated for at least one full day. There were no commercial, agricultural, or industrial activities in the zone; therefore, the legitimate night consumption was estimated based on the recommended assumption that 6% of the population is active and uses water for toilets at the rate of 5 L per flush, as discussed in Fantozzi and Lambert (2012) and Hamilton and McKenzie (2014). Sensitivity analysis of these assumptions are provided in the supplementary data of Chapter 5. After estimating the net night flow in the DMA, the hourly leakage rate during the MNF hour can be found using Equation 2.3. Extrapolating the hourly leakage rate to a daily leakage rate for the normal status in the DMA is only possible when the leakage-pressure relationship is considered. This relationship is considered in the night-day factor (NDF) which was calculated using Equation 2.8. Finally the RL rate was calculated using Equation 2.7. With respect to the Mwanza water supply system, a DMA has been already established in the Kenyatta zone, which contains 64 connections for domestic, commercial, and industrial customers. For this measurement, an ultrasonic water meter with a pressure recorder was installed to measure the flow and pressure at the inlet of the DMA, four pressure recorders were installed at critical points of the DMA, and six recorders were installed at different points in the network to estimate the network pressure. The water was then continuously supplied to the DMA for three consecutive days from 10:45 am on 19 December 2015 to 10:30 am on 21 December 2015. Because this DMA is small, all customer meters in the DMA were read twice at night each day at an interval of two hours to estimate the legitimate night consumption in the DMA. After estimating the net night flow in the DMA, the NDF was calculated and the RL was estimated using Equations 2.8 and 2.7, respectively.

Component analysis of the leakage

The component analysis of the leakage, or the BABE concept, was applied in all the three cases using a spreadsheet model, 'Real Loss Component Analysis: a Tool for Economic Water Loss Control', developed by the Water Research Foundation, USA; (Sturm et al. 2014). The case study data for the reported bursts as well as unreported leaks obtained based on the leakage detection surveys were entered in the model and the leakage was then estimated. All the parameters used in the model are input data of the case studies apart from the flow rate (for which default value was used) as it is not the influencing factor in this analysis (Lambert 1994). Other parameters used in the model were ICF condition factor based on the age of the network; a value of 1 for the N_1 leakage-pressure exponent; and the length of the private line between the customer meter and private boundary of the customer, that is, 0 m, 0 m, and 13.3 m for Sana'a, Zarqa, and Mwanza, respectively.

7.3.1 Results of the water loss component assessment

Figure 7.1 shows the results of the four WL component assessment methods for the three case studies. It shows the IWA standard water balance for the three cases on the left side (in m^3/yr and %). On the right side of Figure 7.1, the breakdown of the WL into AL and RL is shown in a scaled plot. There are two columns: the left one presents the portions of the SIV, BC, and NRW (difference between SIV and BC). If we deduce the unbilled authorised consumption (UAC) from the NRW volume, we obtain the WL volume. Subsequently, we can break down the WL into AL and RL (far-right rectangular column in Figure 7.1). To break down the WL into AL and RL, we use different methods, where each method estimates the AL and RL volumes differently. Thus, the line between AL and RL is drawn differently. For example, based on method 1a (M_{1a}), the WL consists of a very small amount of AL and the remaining part is RL; the line is plotted accordingly (red dotted line M_{1a}). This is similar for the other methods, that is, M_{1b}, M_2, M_3, and M_4. The average of M_{1b} and M_3 is plotted as black line dividing the WL volume into two rectangular parts, where the upper part represents the AL and the lower part presents the RL.

Based on Figure 7.1, the different methods yield different WL components. As expected, the top-down water balance method assuming the UC to be 0.25% of the supplied water (M_{1a}) yields the smallest AL volume in all three cases because the UC in the case studies is considerably higher than this assumption, which is based on a different context in Europe. Therefore, M_{1a} significantly overestimates the RL volume. When increasing the assumption of the UC to 10% of the BC, as in M_{1b}, the method yields relatively higher AL volumes, which are closer to the actual situation of these cases, where UC is common. Nevertheless, it is still unclear how close the results of this method are to the actual AL and RL volumes in the case study systems. The assumption that the UC is 10% of the BC is not based on field data and generalising it for developing countries is not justified because it could significantly differ from one case to another. Based on Figure 7.1, M_{1b} represents the second smallest estimations of the AL in all cases (given that M_3 in Mwanza is not representative for the entire network). Because M_{1b} often estimates a smaller AL than M_2 and M_3, this could indicate that M_{1b} underestimates the AL volume and overestimates the RL volume.

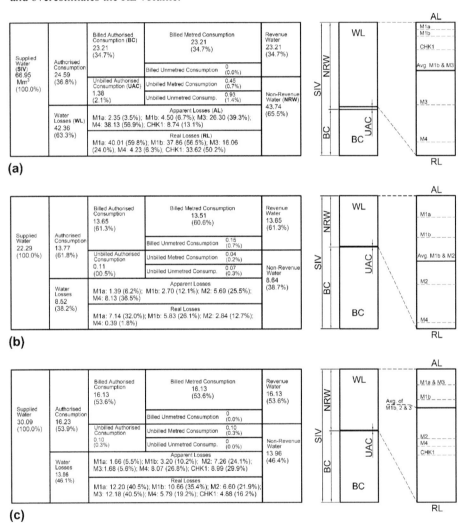

(a)

(b)

(c)

Figure 7.1. : IWA standard water balance in million m^3/yr and % (left) and scaled water loss breakdown (right): (a) Zarqa, Jordan, (b) Sana'a, Yemen, and (c) Mwanza, Tanzania. M_{1a} and M_{1b}: top-down water balance including the unauthorised consumption *of 0.25% of the SIV and 10% of the billed consumption, respectively; M_2: water and wastewater balance method; M_3: minimum night flow analysis method; M_4: component analysis of the leakage method; and CHK1: $M_{1b} - M_4 = M_3$.*

Likewise, the water and wastewater balance method (M_2) probably overestimates the AL volume and therefore underestimates the RL volume because its line is located below the average line in the two cases in Figures 7.1b and c, that is, for Sana'a and Mwanza, respectively. The MNF analysis (M_3) of the two case studies in Figures 7.1a and c for Zarqa and Mwanza are inconsistent because generalizing the RL level of a small area (DMA) for the entire network is associated with significant uncertainties (AL-Washali et al. 2019b). The DMA cannot sufficiently represent the infrastructure, pressure, and consumption of the entire network. The leakage in a DMA in Mwanza with only 64 customers is not representative; the DMA in Zarqa is also not completely representative.

On the other hand, the component analysis of the leakage (BABE; M_4) yields the smallest RL volume in the three cases and therefore overestimates the AL volume. The M_4 estimates the volumes of the bursts that are reported to the utility for repair work, unreported bursts that are discovered based on the leakage detection surveys, and background leaks that cannot be detected by the detection campaigns and continuously run. The records of all burst events in the whole systems were considered in the analysis based on the maintenance records and maintenance software of the utilities. These data are of good quality because burst events were repaired by the utility crew and the maintenance software does not approve or close an administration order for a maintenance team unless technical and geo-referenced data are provided. Therefore, the underestimation of the leakage volume is due to shortcomings of this method with respect to estimating the volume of background leaks and also due to different policies and technologies used for leakage detection in the case studies analysed in this chapter compared with the cases for which this method was initially developed.

A logical check was suggested by Thornton et al. (2008) that the difference of leakage volume of the top-down water balance and the component analysis of leakage (BABE) is closely equal to the leakage volume in the MNF method (CHK1: $M_{1b} - M_4 = M_3$). This check did not yield a close value, neither in Zarqa nor Mwanza, but rather different results, as shown in Figure 7.1.

The comparison of the estimated AL and RL volumes with the actual volumes is impossible, except for one case, that is, if the whole network is divided in DMAs and the flow and pressure of these DMAs are measured throughout the year. This is not the case in the three case studies and will not be the case in the foreseeable future, because of limitations in the capacity and resources of these utilities. Therefore, the results of the WL component assessment methods cannot be validated based on field data. However, the average obtained from the methods M_{1b}, M_2, and M_3 provides a reasonable result in the three cases. It provides closer figures to the subjective expectations of AL and RL portions of SIV, by the specialists in the utilities, which are 25%, 38%; 23%, 15%; and 43%, 11% for Zarqa, Sana'a and Mwanza respectively.

7.4 UNCERTAINTY ANALYSIS

Uncertainty analysis was performed to get insights into the accuracies and sensitivities of the methods as well as the consistency of the methods' outputs. The error propagation theory (Taylor 1997) was used for this analysis and the uncertainties of AL and RL were calculated using the equations presented in Chapter 3. The results of the error propagation theory were also verified with other uncertainty analysis methods, such as variance analysis (Thornton et al. 2008) and Monte Carlo simulation (Rubinstein and Kroese 2016), which provided the same uncertainties.

To estimate the uncertainties in the AL and RL, all components of the standard water balance must be assigned an uncertainty. The supplementary data of this chapter presents the uncertainty in the water balance for all methods and the three case studies. The water balance uncertainties for Zarqa, Sana'a, and Mwanza were assigned based on the estimations and discussions with the specialists of these utilities. As widely applied by IWA WL specialists (Lambert et al. 2014), the uncertainty level is assigned to the water balance component based on the confidence level of the input data. Accordingly, the uncertainty of the SIV was assumed to be ±5% for Zarqa based on the production meter status. The system is almost fully metered and the uncertainty of the BC is thus zero because meter and billing uncertainties are considered for the AL component. The random and systematic errors associated with estimating the amount of the UAC were assumed to be ±5% of the measured volume and ±20% for the unmeasured volume based on the confidence of the specialists of the utility with respect to these figures. The confidence with respect to estimating the data handling errors and inaccuracies of customer meters were also assumed to be ±20% according to the expectations of the specialists in the water utility. The assigned uncertainties were the same when applying different methods for the WL component assessment. These uncertainties are aggregated in the final calculated uncertainties of AL and RL. Two variables remained unassigned with uncertainties, RL and UC; one of them must be assigned an uncertainty and then the uncertainty of the other variable can be calculated depending on the applied WL component assessment method.

For the top-down water balance, the UC must be assigned an arbitrary uncertainty. Therefore, the UC was assigned an uncertainty of ±200% and ±100% for M_{1a} and M_{1b}, respectively, because the assumption of the UC based on these methods does not fit well the case studies. Based on these uncertainties, the aggregated uncertainties of RL were ±9% and ±11% for M_{1a} and M_{1b}, respectively. Similarly, the uncertainties of the water balance were calculated for RL and AL for all methods and the three case studies, as elaborated in the supplementary data of this chapter. Table 7.2 shows the volumes and uncertainties of AL and RL estimated using different methods (M_{1a}, M_{1b}, M_2, M_3, and M_4) for Zarqa, Sana'a, and Mwanza.

Table 7.2. Uncertainties of the AL and RL for the different methods (million m^3)

Method			Zarqa		Sana'a		Mwanza	
			AL	RL	AL	RL	AL	RL
Water Balance	M_{1a}	(Mm³)	2.4	40.0	1.4	7.1	1.7	12.2
		(Δ ±%)	26%	9%	16%	7%	16%	13%
	M_{1b}	(Mm³)	4.5	37.9	2.7	5.8	3.2	10.7
		(Δ ±%)	53%	11%	51%	25%	51%	21%
W&WW Balance	M_2	(Mm³)			5.7	2.8	7.3	6.6
		(Δ ±%)			18%	40%	24%	35%
MNF	M_3	(Mm³)	26.3	16.1			1.7	12.2
		(Δ ±%)	22%	30%			236%	30%
BABE	M_4	(Mm³)	38.1	4.2	8.1	0.4	8.1	5.8
		(Δ ±%)	24%	200%	11%	200%	145%	200%

Regarding the water balance method (M_{1a}), the propagated and aggregated errors of the RL volume are only ±9%, ±7%, and ±13% for Zarqa, Sana'a, and Mwanza, respectively. Similarly, for M_4, the aggregated errors of the AL volume are ±24%, ±11%, and ±145% for Zarqa, Sana'a, and Mwanza, respectively, after assigning an uncertainty level to the RL volume of ± 200% because this method greatly underestimates the RL volume, as discussed in Section 7.3.1.

Interestingly, the methods M_{1a} and M_4 provide the least accurate AL and RL estimations, as discussed above, but they also have relatively low levels of uncertainties (Table 7.2). This suggests that the uncertainty analysis does not indicate how accurate the outputs of the methods or the level of validity of each method are. In fact, for the two WL components (AL and RL), a low level of uncertainty will always be the case when the volume of the final calculated component (e.g. RL) is extremely larger than the volume of the other component (e.g. AL); and high level of uncertainty will always be the case when the volume of the final calculated component is significantly smaller than that of the other component. Further illustrating this fact, Figure 7.2 shows that when the AL is more significant in the network, the aggregated errors of the water balance reach a substantial portion of RL and thus the RL becomes more uncertain through the top down water balance. In contrast, when the AL is insignificant, the aggregated errors of the water balance become less sensitive. In conclusion, it is notable that uncertainty analysis helps in analysing the sensitivities of the inputs of the methods and improving the estimations of the individual methods, but it does not indicate the validity or accuracy of the methods.

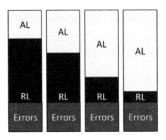

Figure 7.2. : Aggregated errors of the water balance components form a bigger portion of RL when the AL is more significant.

7.5 IMPLICATION OF THE WATER LOSS COMPONENT ASSESSMENT

The impact of the WL component assessment on the leakage reduction planning was analysed in this study using the 'Real Loss Component Analysis: a Tool for Economic Water Loss Control' model (Sturm et al. 2014). The model is a spreadsheet with water balance, cost and financial data and failures and their characteristics as input data. It can be applied to analyse the benefits of different leakage reduction options for continuous and intermittent supply. All the data used in the model are entered except the flow rates of the failures which are not sensitive in the analysis (Lambert 1994). Accordingly, the water and potential monetary savings were calculated for each WL assessment method and three leakage reduction interventions: (i) minimising the response and repair time of bursts in the network, (ii) cost and benefits of conducting leakage detection surveys using acoustic and noise-tracking technologies, and (iii) potential savings based on the reduction of the average pressure of the network.

Minimising the response and repair time of bursts has an influence on the reduction of the runtime of the leakage and is considered in the model. The savings due to the pressure reduction are estimated using the pressure–leakage relationship, which is assumed to be linear (McKenzie et al. 2003; Morrison et al. 2007; Schwaller and van Zyl 2015). The potential savings when conducting regular leakage detection surveys are analysed differently in the model. First, the potentially avoidable and unavoidable leakage volumes are computed using the BABE concept. The frequency of the proactive leakage detection surveys is then estimated using Equations 3.6 and 3.7 (Lambert and Fantozzi 2005). Eventually, the monetary value of these leakage reduction interventions is calculated using the variable cost of the water in each system.

The influence of the WL component assessment on prioritising and planning leakage reduction measures is presented in Figure 7.3. The left side of Figure 7.3 shows the annual cost of WL calculated based on the different methods. The cost of AL differs from the cost of RL. The cost of RL is valued based on the production cost of water, while the cost of AL is the average actual revenue (i.e. price) per cubic metre. Therefore, the total cost of WL varies because the AL and RL volumes and the costs of AL and RL differ from one method to another. Based on the consideration of only M_{1b}, M_2, and M_3 because they yield more reasonable results, as discussed in the previous section, the annual cost of the WL varies from 12.0 to 21.5 million USD for Zarqa, from 3.5 to 4.2 million USD for Sana'a, and from 3.3 to 3.6 million USD for Mwanza. This indicates the sensitivity of the WL component assessment with respect to estimating the cost of the WL and consequently all economic calculations that use this input including the economic level of the WL.

The monetary value of the potential savings of the leakage reduction measures was also analysed for all methods (Figure 7.3, right side). Based on Figure 7.3a and considering only methods with relatively reasonable results, as discussed in the previous section (i.e. M_{1b} and M_3 in Zarqa), the potential savings based on the reduction of the average pressure in Zarqa network by one bar (from 3.3 to 2.2 bar) vary from 1.2 to 2.7 million USD, respectively. The potential savings based on the adoption of the active leakage control (ALC) using regular leakage detection surveys in the entire network every 10.5 months vary from 2.5 to 7.7 million USD. The potential savings based on the reduction of the response and repair time of the reported bursts from the annual average of 2 d to 3 h are 0.4 million USD and are not affected by the component assessment methods because it can only be conducted using M_4, that is, the component analysis of the leakage (BABE). Obviously, the total potential savings based on the adoption of all these measures cannot be the sum of the potential savings of these measures because each option is influenced by other options. For example, the pressure reduction lowers the rate of bursts in the network but undermines the potential of leakage detection surveys because the leakage noise will be harder to detect (AL-Washali et al. 2019b).

Similarly, when considering M_{1b} and M_2 for Sana'a (Figure 7.3b, right side), the potential savings based on the reduction of the average pressure by 0.2 bar (from 1 to 0.8 bar) vary from 0.2 to 0.4 million USD. The potential savings based on the adoption of active leakage control using regular leakage detection surveys in the entire network every 8.9 months vary from 0.7 to 1.7 million USD. The potential savings based on the reduction of the response and repair time of the reported bursts from the annual average of 2.3 to 0.5 day are 0.02 million USD. Based on considering M_{1b}, M_2, and M_3 for Mwanza, the potential savings based on the reduction of the average pressure in the Mwanza network by 2.0 bar (from 5.8 to 3.8 bar) vary from 0.5 to 1.0 million USD. The potential savings based on the adoption of active leakage control using regular leakage detection surveys in the entire network every 10.7 months vary from 0.1 to 1.4 million USD. The potential savings based on the reduction of the response and repair time of the reported bursts from the annual average of 2.0 to 0.5 day are 0.2 million USD.

Based on Figure 7.3, it can be concluded that the feasibility of leakage detection surveys is highly influenced by the component assessment. The feasibility of the pressure management is also influenced but to a lesser extent. The feasibility of the response and repair time reduction is not affected because it was only estimated using one method (BABE). These results confirm that economic planning is significantly affected by the WL component assessment and its uncertainties, leading to unstable and uncertain economic models and WL reduction plans.

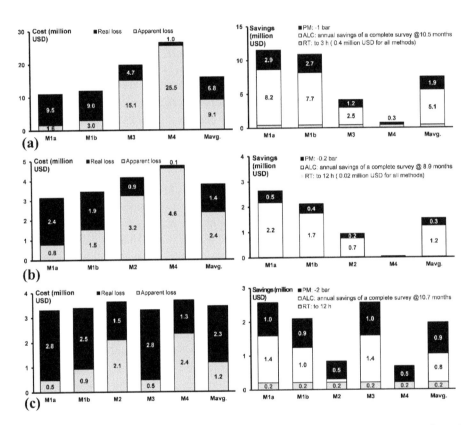

Figure 7.3. : Cost of the water loss components in USD (left) and potential savings based on the reduction of the leak repair time (RT), leak detection surveys (ALC), and pressure management (PM) in USD according to different component assessment methods (right): (a) Zarqa, Jordan, (b) Sana'a, Yemen, and (c) Mwanza, Tanzania.

7.6 CONCLUSIONS

A comparative analysis of the state-of-the-art methods for WL component assessment is presented in this chapter. These methods were applied to three cases in developing countries and economic and uncertainty analyses were performed. The main conclusions of this analysis are the following:

- Improvements of the top-down methods is essential and promising. The top-down water balance will benefit from developing an objective methodology for estimations of the UC volume because the current assumptions are both critical and arbitrary. The accuracy of this method depends on how applicable its assumptions are. In the analysed cases, this method underestimates the AL volume and overestimates the RL volume. Estimating the AL volume using the water and wastewater balance method yields closer results to the expectations of the specialists in these utilities. However, the method could be overestimating the AL volume because it estimates the AL volume more than the average AL volume of all the methods. Applying each method requires verification for the factors and assumptions in each method and their sensitivities and uncertainties. However, if such analysis cannot be carried out, taking the average of these two methods is a practical approach in intermittent supply systems.

- Conducting MNF analysis in one or several DMAs and extrapolating it to the entire network might be justifiable in some cases, but it is not very rational because every DMA differs in terms of the mains length, service connections, pressure, and burst frequencies. The MNF analysis is more suitable for the DMA-scale than for the system-wide scale with respect to the interventions, identification, and repair of unreported leaks. The component analysis of the leakage (BABE) method remains the only way to break down the leakage into subcomponents, enabling the water utilities to understand the nature and behaviour of the leakage in their systems. However, the BABE analyses only a small portion of the leakage and cannot be used for WL component assessment.

- The results show that WL component assessment has significant uncertainties, which in turn affect the cost of WL and substantially impact the planning of RL and AL minimisation measures. Addressing this issue needs more investigation on how the WL component assessment can be improved. Field observations that could help to validate and calibrate the methods are not obtainable unless the entire network is divided into DMAs to conduct regular MNF measurements throughout the year, which is very costly and unlikely, especially in developing countries. On the other hand, the uncertainty analysis helps to improve the output of the individual methods but not the accuracies of the methods. Therefore, assessing the WL components by using at least two methods should improve the prioritisation, economic modelling, monitoring, and benchmarking of the WL.

- For intermittent supply systems in developing countries, the average volume of the AL from the top-down water balance and water and wastewater balance methods should be used to establish the standard water balance. The RL can then be further broken down using BABE. Based on this methodology, leakage reduction interventions can be planned and prioritised for the entire network. Subsequently, MNF analysis can be used on a DMA-scale in the implementation phase to separately intervene, monitor, and reduce the leakage in each DMA.

7.7 SUPPLEMENTARY DATA

7.7.1 Uncertainties of the water balance in different methods for the three case studies

Table 7.3: Uncertainty analysis for different water loss component assessment methods in Zarqa

Component		M1a (m³)	U(±%)	M1b (m³)	U(±%)	M3 (m³)	U(±%)	M4 (m³)	U(±%)
SIV		66,951,078	5%						
BC	BMC	23,208,261	0%						
	BUC	0	0%						
NRW		43,742,817	8%						
UAC	UMC	447,020	5%						
	UUC	933,316	20%						
		1,380,336	14%						
WL		42,362,481	8%						
AL	UC	167,378	200%	2,320,826	100%	24,121,824	24%	35,945,454	25%
	CMIs	2,570,993	20%	2,570,993	20%	2,570,993	20%	2,570,993	20%
	DHEs	-388,226	20%	-388,226	20%	-388,226	20%	-388,226	20%
		2,350,145	26%	4,503,593	53%	26,304,591	22%	38,128,221	24%
RL		40,012,336	9%	37,858,888	11%	16,057,890	30%	4,234,260	200%

(Za)

Table 7.4: Uncertainty analysis for different water loss component assessment methods in Sana'a

Component		M1a (m³)	U(±%)	M1b (m³)	U(±%)	M2 (m³)	U(±%)	M4 (m³)	U(±%)
SIV		22,290,314	2%						
BC	BMC	13,506,402	0%						
	BUC	146,220	10%						
NRW		8,637,692	5%						
UAC	UMC	44,262	5%						
	UUC	69,890	33%						
		114,152	20%						
WL		8,523,540	5%						
AL	UC	57,076	200%	1,366,613	100%	4,356,071	24%	6,802,259	14%
	CMIs	560,516	20%	560,516	20%	560,516	20%	560,516	20%
	DHEs	769,865	20%	769,865	20%	769,865	20%	769,865	20%
		1,387,457	16%	2,696,993	51%	5,686,452	18%	8,132,640	11%
RL		7,136,083	7%	5,826,547	25%	2,837,088	40%	390,900	200%

(Su)

Table 7.5: Uncertainty analysis for different water loss component assessment methods in Mwanza

Component		M1a (m³)	U(±%)	M1b (m³)	U(±%)	M2 (m³)	U(±%)	M3 (m³)	U(±%)	M4 (m³)	U(±%)
SIV		30,092,916	5%								
BC	BMC	16,130,168	0%								
	BUC	0	0%								
NRW		13,962,748	11%								
UAC	UMC	101,189	10%								
	UUC	0	0%								
		101,189	11%								
WL		13,861,559	11%								
AL	UC	75,232	200%	1,613,017	100%	5,676,647	30%	93,549	423%	6,489,317	182%
	CMIs	774,248	20%	774,248	20%	774,248	20%	774,248	20%	774,248	20%
	DHEs	809,734	20%	809,734	20%	809,734	20%	809,734	20%	809,734	20%
		1,659,214	16%	3,196,999	16%	7,260,629	24%	1,677,531	236%	8,073,399	145%
RL		12,202,345	13%	10,664,560	21%	6,600,930	35%	12,184,029	30%	5,788,260	200%

(Mw)

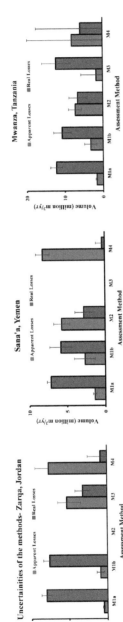

Figure 7.4. : Uncertainties of the different methods in Zarqa, Sana'a and Mwanza.

Acronyms

M1a	Top-Down Water Balance with assuming unauthorised consumption at 0.25%SIV
M1b	Top-Down Water Balance with assuming unauthorised consumption at 10%BC
M2	Water and Wastewater Balance
M3	MNF analysis
M4	BABE
Za	Zarqa, Jordan
Sa	Sana'a, Yemen
Mw	Mwanza, Tanzania
SIV	System Input Volume
BC	Billed Consumption
BMC	Billed Metered Consumption
BUC	Billed Unmetered Consumption
NRW	Non-Revenue Water
UAC	Unbilled Authorised Consumption
UMC	Unbilled Metered Authorised Consumption
UUC	Unbilled Unmetered Authorised Consumption
WL	Water Loss
UC	Unauthorised Consumption
CMIs	Customer Meter Inaccuracies
DHEs	Data Handling Errors
AL	Apparent Loss
RL	Real Loss

7.7.2 Summary of model outputs: benefits of different reduction options for the case studies

Zarqa, Jordan

WaterRF 4372: Real Loss Component Analysis: A Tool for Economic Water Loss Control

Water Audit: Zarqa Water Utility, AL-Zarqa, Jordan, 2014
WaterRF 4372 COMPONENT ANALYSIS MODEL SUMMARY

As the input data is filled into the model, this sheet will populate with the results and recommendations from the Real Losses Component Analysis, A-L-R Times, Economic Intervention and Pressure Management tabs. The performance indicators from the AWWA Free Water Audit Software have been added to show a brief review of the performance of the system being analyzed

WATER AUDIT PERFORMANCE INDICATORS

Financial

Non-revenue water as percent by volume of water supplied:	65.3%	
Non-revenue water as percent by cost of operating system:	59.9%	
Annual cost of Apparent Losses:	$10,320,742	
Annual cost of Real Losses:	$6,389,138	

Operational Efficiency

Apparent Losses per service connection per day:	485.1	litres/service conn/day
Real Losses per service connection per day*:	848.9	litres/service conn/day
Real Losses per length of main per day:	N/A	litres/km/day
Real Losses per service connection per day per metres (head) pressure:	25.7	litres/service conn/day/mH
Unavoidable Annual Real Losses (UARL):	1,324.42	ML/Yr
Current Annual Real Losses (CARL):	26,958.39	ML/Yr
Infrastructure Leakage Index (ILI) [CARL/UARL]:	20.4	

REAL LOSS COMPONENT ANALYSIS RESULTS

System Component	Background Leakage	Reported Failures	Unreported Failures	Total
	(ML)	(ML)	(ML)	(ML)
Reservoirs	-	-	-	-
Mains and Appurtenances	530.80	1,223.51	12.06	**1,766.36**
Service Connections	1,287.34	1,158.57	22.00	**2,467.90**
Total Annual Real Loss	**1,818.13**	**2,382.08**	**34.06**	4,234.26
		Real Losses as Calculated by Water Audit		26,958.39
		Hidden Losses/Unreported Leakage Currently Running Undetected		22,724.13

AWARNESS, LOCATION AND REPAIR TIME REDUCTION RESULTS

	Reported Failures	Unreported Failures	
Total Potential Savings if Location and Repair Duration is Reduced as Simulated on the A-L-R Times Options Sheet	1,850.8	21.3	(ML)
Total Potential Cost Savings if Location and Repair Duration is Reduced as Simulated on the A-L-R Times Options Sheet	$ 438,651	$ 5,045	Per Year

ECONOMIC INTERVENTION FREQUENCY FOR PROACTIVE LEAK DETECTION RESULTS

Percentage of the System to be Surveyed per Year	114	%
Average Annual Budget for Intervention (Proactive Leak Detection)	255,395	$/year
Potentially Recoverable Leakage	21,646.51	ML/year

ALTERNATIVE PRESSURE MANAGEMENT SCENARIO RESULTS

User-Inputted Reduction in Average System Pressure	10.0	metres (head)
Assumed % Reduction in Average System Pressure	30%	
Estimated Real Loss Reduction from Pressure Management Program	8,169.2	MG/Yr
Financial Savings from Pressure Management Program	1,936,102	$/Year
User-Estimated Cost of Pressure Reduction	1,000,000	$
Resulting Pressure Management Program Payback Period	0.5	Years

September, 2018   ©2014 Water Research Foundation. ALL RIGHTS RESERVED.

Figure 7.5. : Summary of model outputs: benefits of different reduction options for Zarqa, Jordan.

Water Audit: Zarqa Water Utility, AL-Zarqa, Jordan, 2014
CALCULATION OF ECONOMIC INTERVENTION FREQUENCY FOR PROACTIVE LEAK DETECTION

System Characteristics		
Total Length of Mains	2,242.0	kilometers
Number of Service Connections	87,000	service connections
Service Connection Density	38.8	conn./kilometer main
Average System Pressure	33.0	metres (head)

	Water Balance Results		
TBL	Current Annual Background Leakage	1,818.13	ML/Yr
CRL	Real Losses from Current Reported Leakage	2,382.08	ML/Yr
UL	Unreported Failures Identified Through Existing Proactive Leak Detection Program	34.06	ML/Yr
	Hidden Failures/Unreported Failures not Identified or Captured by Current Leakage Management Policy	22,724.13	ML/Yr
CARL	Current Annual Real Losses	26,958.39	ML/Yr
UARL	Unavoidable Annual Real Losses	1,324.42	ML/Yr
	Infrastructure Leakage Index (ILI) [CARL/UARL]	20.4	

	Variable Cost of Real Losses		
CV	Variable Production cost (applied to Real Losses)	0.24	$/m3
		237.00	$/ML
CI	Cost of comprehensive leak detection survey (excluding leak repair cost)	100.00	$/km
		224,200	$/for entire system
RR	Average Rate of Rise of Unreported Leakage	3.00	m3/km of mains/day in a year
		6.73	ML/day in a year
	CI/CV	421.9	m3/km
EIF	Economic Intervention Frequency [0.789 * (CI/CV)/(RR)] ^0.5	10.5	months
		320.4	days
	Economic Intervention Frequency - Average Leak Run Time	160.2	days
	Economic Percentage of System to be Surveyed per Year	114	%
ABI	Average Annual Budget for Intervention (Proactive Leak Detection)	255,395	$/year
EUL	Economic Unreported Real Losses	1,077,618	m3/year
		1,077.6	ML/year
	Economic Infrastructure Leakage Index (ILI)	4.0	
PRL	Potentially Recoverable Leakage (CARL-CRL-EUL-TBL-UL)	21,646.5	ML/year

September 2015

Instructions: Use this sheet to establish a preliminary schedule for proactive leak detection surveys and the corresponding necessary budget. Once the results from consecutive leak surveys are available, the Rate of Rise of Unreported Leakage should be updated and the proactive leak detection schedule should be refined taking into consideration these findings.

In order to establish a preliminary schedule for proactive leak detection or to review the currently utilized proactive leak detection schedule enter the cost for undertaking proactive leak detection ($/mile or $/km) in cell D31. Next enter the Average Rate of Rise of Unreported Leakage in cell D34. The Average Rate of Rise of Unreported Leakage is the rate at which leakage increases with time. The rate of rise is not necessarily linear since it can quickly change due to seasonal effects and other system specific impacts. The AWWA M36 Manual recommends assessing The Average Rate of Rise of Unreported Leakage either by comparing water balance results of several consecutive years and calculating the Average Rate of Rise of Unreported Leakage based on the increase in Real Loss volume from year to year (if utility does not employ proactive leak detection), or by analysis of District Metered Area might flow data or repair records of leaks detected through proactive leak detection (if utility employs proactive leakage control). Further details about how to assess the Average Rate of Rise of Unreported Leakage are provided in the AWWA M36 Manual. Note: kgal = 1,000 gallons.

Please note that the calculations used to determine the Economic Intervention Frequency are not best fit for systems with infrastructure Leakage Index values that approach 1. For such systems, it is advised to review the Economic Intervention Frequency and associated costs

Current Annual Real Losses vs. Potentially Recoverable Leakage Through Proactive Leak Detection

■ Current Annual Real Losses
■ Potentially Recoverable Leakage Through Proactive Leak Detection

26,958.39   21,646.51

Figure 7.6. : Summary of model parameters and output of leakage detection potential for Zarqa, Jordan.

Sana'a, Yemen

WaterRF 4372: Real Loss Component Analysis: A Tool for Economic Water Loss Control

Water Audit: Sana'a Water Supply and Sanitation Local Corporation, Sana'a, Yemen, 2016
WaterRF 4372 COMPONENT ANALYSIS MODEL SUMMARY

As the input data is filled into the model, this sheet will populate with the results and recommendations from the Real Losses Component Analysis, A-L-R Times, Economic Intervention and Pressure Management tabs. The performance indicators from the AWWA Free Water Audit Software have been added to show a brief review of the performance of the system being analyzed

WATER AUDIT PERFORMANCE INDICATORS

Financial		
Non-revenue water as percent by volume of water supplied:	38.8%	
Non-revenue water as percent by cost of operating system:	26.4%	
Annual cost of Apparent Losses:	$2,389,282	
Annual cost of Real Losses:	$1,430,817	

Operational Efficiency		
Apparent Losses per service connection per day:	128.8	litres/service conn/day
Real Losses per service connection per day*:	133.1	litres/service conn/day
Real Losses per length of main per day:	N/A	litres/km/day
Real Losses per service connection per day per metres (head) pressure:	13.3	litres/service conn/day/mH
Unavoidable Annual Real Losses (UARL):	324.14	ML/Yr
Current Annual Real Losses (CARL):	4,331.82	ML/Yr
Infrastructure Leakage Index (ILI) [CARL/UARL]:	13.4	

REAL LOSS COMPONENT ANALYSIS RESULTS

System Component	Background Leakage	Reported Failures	Unreported Failures	Total
	(ML)	(ML)	(ML)	(ML)
Reservoirs	-	-	-	-
Mains and Appurtenances	22.99	23.62	-	46.61
Service Connections	132.07	45.73	-	177.81
Total Annual Real Loss	155.06	69.35	-	224.41
Real Losses as Calculated by Water Audit				4,331.82
Hidden Losses/Unreported Leakage Currently Running Undetected				4,107.40

AWARNESS, LOCATION AND REPAIR TIME REDUCTION RESULTS

	Reported Failures	Unreported Failures	
Total Potential Savings if Location and Repair Duration is Reduced as Simulated on the A-L-R Times Options Sheet	52.8	-	(ML)
Total Potential Cost Savings if Location and Repair Duration is Reduced as Simulated on the A-L-R Times Options Sheet	$ 17,449	$ -	Per Year

ECONOMIC INTERVENTION FREQUENCY FOR PROACTIVE LEAK DETECTION RESULTS

Percentage of the System to be Surveyed per Year	134	%
Average Annual Budget for Intervention (Proactive Leak Detection)	130,110	$/year
Potentially Recoverable Leakage	3,713.49	ML/year

ALTERNATIVE PRESSURE MANAGEMENT SCENARIO RESULTS

User-Inputted Reduction in Average System Pressure	2.0	metres (head)
Assumed % Reduction in Average System Pressure	20%	
Estimated Real Loss Reduction from Pressure Management Program	866.4	MG/Yr
Financial Savings from Pressure Management Program	286,163	$/Year
User-Estimated Cost of Pressure Reduction	-	$
Resulting Pressure Management Program Payback Period	-	Years

September, 2018 ©2014 Water Research Foundation. ALL RIGHTS RESERVED.

Figure 7.7. : Summary of model outputs: benefits of different reduction options for Sana'a, Yemen.

WaterRF 4372: Real Loss Component Analysis: A Tool for Economic Water Loss Control

Water Audit: Sana'a Water Supply and Sanitation Local Corporation, Sana'a, Yemen, 2016
CALCULATION OF ECONOMIC INTERVENTION FREQUENCY FOR PROACTIVE LEAK DETECTION

		Value to be entered by the user
		Value is automatically filled in/calculated by Model
		Recommended default value

System Characteristics

Total Length of Mains	967.5	kilometers
Number of Service Connections	88,908	service connections
Service Connection Density	91.9	conn./kilometer main
Average System Pressure	10.0	(metres (head)

Water Balance Results

TBL	Current Annual Background Leakage	155.06	ML/Yr
CRL	Real Losses from Current Reported Leakage	69.35	ML/Yr
UL	Unreported Failures Identified Through Existing Proactive Leak Detection Program	-	ML/Yr
	Hidden Failures/Unreported Failures not Identified or Captured by Current Leakage Management Policy	4,107.40	ML/Yr
CARL	Current Annual Real Losses	4,331.82	ML/Yr
UARL	Unavoidable Annual Real Losses	324.14	ML/Yr
	Infrastructure Leakage Index (ILI) [CARL/UARL]	13.4	

Variable Cost of Real Losses

CV	Variable Production cost (applied to Real Losses)	0.33	$/m3
		330.30	$/ML
CI	Cost of comprehensive leak detection survey (excluding leak repair cost)	100.00	$/km
		96,750	$/ for entire system
RR	Average Rate of Rise of Unreported Leakage	3.00	m3/km of mains/day in a year
		2.90	ML/day in a year
	CI/CV	302.8	m3/km
EIF	Economic Intervention Frequency [0.789 * (CI/CV)/RR] ^ 0.5	8.9	months
		271.4	days
	Economic Intervention Frequency - Average Leak Run Time	135.7	days
	Economic Percentage of System to be Surveyed per Year	134	%
ABI	Average Annual Budget for Intervention (Proactive Leak Detection)	130,110	$/year
EUL	Economic Unreported Real Losses	393,910	m3/year
		393.9	ML/year
	Economic Infrastructure Leakage Index (ILI)	1.9	
PRL	Potentially Recoverable Leakage [CARL-CRL-EUL-TBL-UL]	3,713.5	ML/year

September, 2016

Instructions: Use this sheet to establish a preliminary schedule for proactive leak detection surveys and the corresponding necessary budget. Once the results from consecutive leak surveys are available, the Rate of Rise of Unreported Leakage should be updated and the proactive leak detection schedule should be refined taking into consideration these findings.

In order to establish a preliminary schedule for proactive leak detection or to review the currently utilized proactive leak detection schedule enter the cost for undertaking proactive leak detection ($/mile or $/km) in cell D31. Next enter the Average Rate of Rise of Unreported Leakage in cell D34. The Average Rate of Rise of Unreported Leakage is the rate at which leakage increases with time. The rate of rise is not necessarily linear since it can quickly change due to seasonal effects and other system specific impacts. The AWWA M36 Manual recommends assessing The Average Rate of Rise of Unreported Leakage either by comparing water balance results of several consecutive years and calculating the Average Rate of Rise of Unreported Leakage based on the increase in Real Loss volume from year to year (if utility does not employ proactive leak detection) or by analysis of District Metered Area night flow data or repair records of leaks detected through proactive leak detection (if utility employs proactive leakage control). Further details about how to assess the Average Rate of Rise of Unreported Leakage are provided in the AWWA M36 Manual. Note: kgal = 1,000 gallons.

Please note that the calculations used to determine the Economic Intervention Frequency are not best fit for systems with infrastructure Leakage Index values that approach 1. For such systems, it is advised to review the Economic Intervention Frequency and associated costs.

Current Annual Real Losses vs. Potentially Recoverable Leakage Through Proactive Leak Detection

Figure 7.8.: Summary of model parameters and output of leakage detection potential for Sana'a, Yemen.

Mwanza, Tanzania

WaterRF 4372: Real Loss Component Analysis: A Tool for Economic Water Loss Control

Water Audit: Mwanza Water Utility, Mwanza, Tanzania, 2015
WaterRF 4372 COMPONENT ANALYSIS MODEL SUMMARY

As the input data is filled into the model, this sheet will populate with the results and recommendations from the Real Losses Component Analysis, A-L-R Times, Economic Intervention and Pressure Management tabs. The performance indicators from the AWWA Free Water Audit Software have been added to show a brief review of the performance of the system being analyzed

WATER AUDIT PERFORMANCE INDICATORS

Financial		
Non-revenue water as percent by volume of water supplied:	46.4%	
Non-revenue water as percent by cost of operating system:	43.4%	
Annual cost of Apparent Losses:	$1,178,842	
Annual cost of Real Losses:	$2,269,370	

Operational Efficiency		
Apparent Losses per service connection per day*:	224.9	litres/service conn/day
Real Losses per service connection per day*:	545.7	litres/service conn/day
Real Losses per length of main per day:	N/A	litres/km/day
Real Losses per service connection per day per metres (head) pressure:	9.4	litres/service conn/day/mH
Unavoidable Annual Real Losses (UARL):	1,513.11	ML/Yr
Current Annual Real Losses (CARL):	9,816.51	ML/Yr
Infrastructure Leakage Index (ILI) [CARL/UARL]:	6.5	

REAL LOSS COMPONENT ANALYSIS RESULTS

System Component	Background Leakage	Reported Failures	Unreported Failures	Total
	(ML)	(ML)	(ML)	(ML)
Reservoirs	-	-	-	-
Mains and Appurtenances	191.97	1,247.75	1,114.63	**2,554.35**
Service Connections	920.75	1,484.60	828.56	**3,233.91**
Total Annual Real Loss	1,112.72	2,732.36	1,943.18	5,788.26
Real Losses as Calculated by Water Audit				9,816.51
Hidden Losses/Unreported Leakage Currently Running Undetected				4,028.24

AWARNESS, LOCATION AND REPAIR TIME REDUCTION RESULTS

	Reported Failures	Unreported Failures	
Total Potential Savings if Location and Repair Duration is Reduced as Simulated on the A-L-R Times Options Sheet	903.4	34.2	(ML)
Total Potential Cost Savings if Location and Repair Duration is Reduced as Simulated on the A-L-R Times Options Sheet	$ 208,857	$ 7,908	Per Year

ECONOMIC INTERVENTION FREQUENCY FOR PROACTIVE LEAK DETECTION RESULTS

Percentage of the System to be Surveyed per Year	113	%
Average Annual Budget for Intervention (Proactive Leak Detection)	97,881	$/year
Potentially Recoverable Leakage	3,604.84	ML/year

ALTERNATIVE PRESSURE MANAGEMENT SCENARIO RESULTS

User-Inputted Reduction in Average System Pressure	20.0	metres (head)
Assumed % Reduction in Average System Pressure	34%	
Estimated Real Loss Reduction from Pressure Management Program	3,385.0	MG/Yr
Financial Savings from Pressure Management Program	782,541	$/Year
User-Estimated Cost of Pressure Reduction	-	$
Resulting Pressure Management Program Payback Period	-	Years

Figure 7.9. : Summary of model outputs: benefits of different reduction options for Mwanza, Tanzania.

WaterRF 4372: Real Loss Component Analysis: A Tool for Economic Water Loss Control

Water Audit: Mwanza Water Utility, Mwanza, Tanzania 2015
CALCULATION OF ECONOMIC INTERVENTION FREQUENCY FOR PROACTIVE LEAK DETECTION

Instructions: Use this sheet to establish a preliminary schedule for proactive leak detection surveys and the corresponding necessary budget. Once the results from consecutive leak surveys are available, the Rate of Rise of Unreported Leakage should be updated and the proactive leak detection schedule should be refined taking into consideration these findings.

In order to establish a preliminary schedule for proactive leak detection or to review the currently utilized proactive leak detection schedule enter the cost for undertaking proactive leak detection ($/mile or $/km) in cell D31. Next enter the Average Rate of Rise of Unreported Leakage in cell D34. The Average Rate of Rise of Unreported Leakage is the rate at which leakage increases with time. The rate of rise is not necessarily linear since it can quickly change due to seasonal effects and other system specific impacts. The AWWA M36 Manual recommends assessing The Average Rate of Rise of Unreported Leakage either by comparing water balance results of several consecutive years and calculating the Average Rate of Rise of Unreported Leakage based on the increase in Real Loss volume from year to year (if utility does not employ proactive leak detection), or by analysis of District Metered Area night flow data or repair records of leaks detected through proactive leak detection (if utility employs proactive leakage control). Further details about how to assess the Average Rate of Rise of Unreported Leakage are provided in the AWWA M36 Manual. Note: kgal = 1,000 gallons.

Please note that the calculations used to determine the Economic Intervention Frequency are not best fit for systems with infrastructure Leakage Index values that approach 1. For such systems, it is advised to review the Economic Intervention Frequency and associated costs

System Characteristics		
Total Length of Mains	870.0	kilometers
Number of Service Connections	49,284	service connections
Service Connection Density	56.6	conn./kilometer main
Average System Pressure	58.0	metres (head)

Water Balance Results			
TBL	Current Annual Background Leakage	1,112.72	ML/Yr
CRL	Real Losses from Current Reported Leakage	2,732.36	ML/Yr
UL	Unreported Failures Identified Through Existing Proactive Leak Detection Program	1,943.18	ML/Yr
	Hidden Failures/Unreported Failures not Identified or Captured by Current Leakage Management Policy	4,028.24	ML/Yr

CARL	Current Annual Real Losses	9,816.51	ML/Yr
UARL	Unavoidable Annual Real Losses	1,513.11	ML/Yr
	Infrastructure Leakage Index (ILI) [CARL/UARL]	6.5	

	Variable Cost of Real Losses		
CV	Variable Production cost (applied to Real Losses)	0.23	$/m3
		231.18	$/ML
CI	Cost of comprehensive leak detection survey (excluding leak repair cost)	100.00	$/km
		87,000.00	$/for entire system
RR	Average Rate of Rise of Unreported Leakage	3.00	m3/km of mains/day in a year
		2.61	ML/day in a year
	CI/CV	432.6	m3/km
EIF	Economic Intervention Frequency (0.789 * (CI/CV)/RR) ^ 0.5	10.7	months
		324.4	days
	Economic Intervention Frequency - Average Leak Run Time	162.2	days
	Economic Percentage of System to be Surveyed per Year	113	%
ABI	Average Annual Budget for Intervention (Proactive Leak Detection)	97,881	$/year
EUL	Economic Unreported Real Losses	423.398	m3/year
		423.4	ML/year
	Economic Infrastructure Leakage Index (ILI)	4.1	
PRL	Potentially Recoverable Leakage (CARL-CRL-EUL-TBL-UL)	3,604.8	ML/year

Current Annual Real Losses vs. Potentially Recoverable Leakage Through Proactive Leak Detection

9,816.51

3,604.84

■ Current Annual Real Losses
■ Potentially Recoverable Leakage Through Proactive Leak Detection

September 2015

Figure 7.10. : Summary of model parameters and output of leakage detection potential for Mwanza, Tanzania

8

IMPACT OF FLOAT-VALVES ON WATER METER PERFORMANCE UNDER INTERMITTENT AND CONTINUOUS SUPPLY

Intermittent supply is common worldwide. It triggers households with piped connection to adjust the supply scheme by the use of a water tank with a float valve (FV) at the entrance, which has a major influence on the water meter accuracy. This chapter investigated the impact of the water tank with a FV on the performance of water meters under intermittent and continuous supply conditions, using laboratory experiments, field measurements, and hydraulic modelling. Results revealed that the inflows into the water tank are consistently lower than the outflows of the tank. This will always be the case owing to the balancing mechanism of the tank. The flows that pass through the water meter represent the inflows into the tank. Therefore, higher metering errors and more apparent losses are expected for a combination of a water tank, FV, and continuous supply. Besides, different FV types have different hydraulic characteristics. Larger FVs with higher discharge rates tend to maintain the water level close to the full level in the tank and conferred longer periods of low flows, worse meter performance, and more apparent losses. For intermittent supply, results confirmed that higher intermittency levels leads to improved performance of water meters and reduce the apparent losses. This points to the complication in transformation from intermittent to continuous supply worldwide. In this case, water utilities should expect higher meter errors and more revenue losses unless the meter replacement policy recognises lower flows passing through the meter.

This chapter has been published as: AL-Washali, T., Mahardani, M., Sharma, S., Arregui, F., and Kennedy, M. "Impact of Float-Valves on Customer Meter Performance under Intermittent and Continuous Supply Conditions." Resources, Conservation and Recycling, 163, 105091, 2020.

8.1 INTRODUCTION

The water meter is a cash register, a system management tool, and a conservation instrument. Yet, water meters are not absolutely accurate measuring instruments. All water meters, including new ones, have drawbacks (Arregui et al. 2015). Varied measuring limitations exist for different meters, depending on the metering technology and the meter class (Arregui et al. 2006a; Van Zyl 2011). Mechanical meters are commonly used to measure the water consumption for customers of water utilities. They are either volumetric meters that measure pockets of water directly such as rotating piston meters, or inferential (or velocity) meters that infer the volumetric flow rate from the velocity of the water, such as Woltmann, Single, and Multi-Jet meters. Electromagnetic and ultrasonic meters are marginal technologies that are used in limited, specific cases because of their high cost and power requirements. They detect water velocity using electromagnetic principles and ultrasound waves. To evaluate the performance of new meters, several standards and guidelines exist (ISO 2014a; ISO 2014b; OIML 2013a; OIML 2013b). Each new meter should be tested at four main flows (ISO 2014a; OIML 2013a). The minimum flow rate (q_1 or q_{min}) is the lowest flow rate at which the meter is required to operate within the maximum permissible error (MPE). The transitional flow rate (q_2 or q_t) is the flow between the minimum flow rate and the nominal flow rate. The permanent flow rate (q_3 or q_p) is the highest flow rate within the rated operating conditions at which the meter operates within the MPE. Finally, the overloaded flow rate (q_4 or q_{max}) is the highest flow rate at which the meter is required to operate for a short period of time within the MPE while maintaining its metrological performance. While q_p and q_{max} depend on the size of the meter, different classes of meters have different ratios of q_t and q_{min} to q_p (Figure 8.1). The required accuracy tolerance, or MPE, for new meters is ±5% in the lower zone and ±2% in the upper zone of Figure 8.1. Depending on the water utility's policy, the accuracy tolerance of the used meters can be doubled in some guidelines (Orden_ICT_155 2020) or widened to ±8% and ±3.5% for the lower and upper zones, respectively (Ncube and Taigbenu 2019; Van Zyl 2011; Walter et al. 2018).

Recommendations on the selection of the appropriate meter type are proposed based on the consumption data (Johnson 2001) and criteria that include low flow accuracy, ability to pass particulates, and accuracy degradation rate (Mutikanga 2014).

Once the meter is installed in the field, its performance starts to decline. The field meter accuracy is determined by several factors (Arregui et al. 2005; Criminisi et al. 2009; Thornton and Rizzo 2002) including meter wear and tear, blockage of the meter inlet or strainer, depositions of the meter components, incorrect sizing, incorrect mounting position, and incorrect flow profile. While volumetric meters such as the oscillating piston and nutating disc are sensitive to water quality and suspended particles, velocity meters such as single and multiple jet meters are more sensitive to low flows and drag torque on the sensor element.

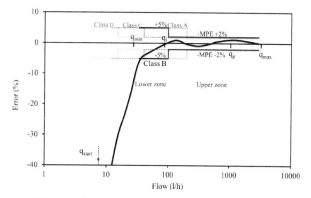

Figure 8.1. Water meter error curve parameters

There are ample studies on the field performance of the meter (Arregui et al. 2006a; Arregui et al. 2018b; Arregui et al. 2006b; Couvelis and Van Zyl 2015; Mantilla-Peña et al. 2018; Moahloli et al. 2019; Mutikanga et al. 2011a; Ncube and Taigbenu 2019; Stoker et al. 2012; Walter et al. 2018; Yazdandoost and Izadi 2018). The inaccuracy level of the used meters depends on the metrological performance of the meter at each flow rate. The share of the flow rate that passes through the meter is, therefore, a governing factor of the meter accuracy (Arregui et al. 2006a; Fontanazza et al. 2015; Male et al. 1985). Obviously, the meter performance at low flows is critical (Arregui et al. 2015; De Marchis et al. 2014; Richards et al. 2010). The error curve does not start from the axis of ordinates but from a threshold on the x-axis before which the flows are not registered by the meter, because friction forces prevent the meter sensor from moving. After the starting flow, the performance of the meter starts to improve, with a rather low accuracy, till q_{min}, when the meter begins to operate within the MPE.

Of the total flow passing through the meter, the higher the proportion of the low flows, the higher the meter inaccuracy. This shows the effect of the consumption profile on the meter accuracy. For this reason, the weighted error of the meter is proposed (Arregui et al. 2006a; Shields et al. 2012) to relate the demand consumption flows to the error level of the meter. The weighted error considers the consumption flows that pass through the meter, therefore, it is a good indicator of the meter's field metrological performance (Arregui et al. 2018b).

However, this should not be the case in intermittent supply where a water tank with an attached float-valve (FV) exists (De Marchis et al. 2014; De Marchis et al. 2015; Tamari and Ploquet 2012). Intermittent supply is common worldwide (Charalambous and Laspidou 2017; De Marchis et al. 2010). In low and middle income countries, the population with piped water on premises is 3.2 billion, of which 1.3 billion people receive intermittent supply (Charalambous 2019). A network affected by intermittency is stressed by repetitive pressure transients (with entrapped air), causing network deterioration, more leaks and breaks and contamination risks. In an intermittent supply regime, depending upon the location of the water meter and water tanks, the meter accuracy is affected by the iterative filling and emptying process of the water tank.

Rizzo and Cilia (2005) tested and compared the accuracy of a meter installed at the inlet of a water tank with a meter installed at the outlet of the water tank associated with a FV, and found that the inlet meter constantly under-recorded between 5 and 9% of the water measured by the outlet meter. This was due to the effect of the FV and the filling process in the tank. Criminisi et al. (2009) developed a mathematical model to simulate the same arrangement of water tank, FV, water meter, and user consumption. The tank filling process depends on the network pressure, FV characteristics, and tank water level. However, some of the model parameters required laboratory characterisation and cannot be generalised for all systems. De Marchis et al. (2013) implemented a mathematical model to assess apparent losses caused by meter under-registration based on a hydraulic network model, a pressure-reducing valve model, a pressure-driven demand, and an apparent losses model, considering the complexity of private tanks yet with constant valve characteristics. The combined influence of pressure reduction and tank filling on the meter error was highlighted by De Marchis et al. (2014) who implemented the model developed by Criminisi et al. (2009) and found that in a specific FV and tank arrangement, apparent losses can be generally over-registration for intermittently operated networks and under-registration for tanks that are almost full. The study recommended more investigation with other FV and tank characteristics. This chapter presents the results of a study carried out to investigate the impacts of different water supply intermittency and continuity conditions and different tank and FV characteristics on the customer meter performance and level of apparent losses. The influence of intermittent supply level on the apparent loss level was investigated and impact of transforming from intermittent to continuous supply was highlighted for customers whose supply system contains a water tank and FV arrangement. The effects of the characteristics of different FVs and tank sizes were also analysed. The results of this study will aid water utilities to understand the impact of critical factors affecting apparent losses and will assist them in managing apparent losses in distribution networks where intermittent supply and private water tanks with FVs are common.

8.2 RESEARCH METHODOLOGY

In principle, the meter accuracy is affected by the FV flow rate, which is affected by the water level in the tank, which is affected by the water consumption during the day. Therefore, the methodology of this study started by first defining the meter error—flow relationship, based on bench test experiments for a sample of used meters. Second, the hydraulic characteristics of the FV were experimented to obtain the FV resistance coefficient (K) for each corresponding water level in the tank (h). Non-linear regression analysis was used to empirically model the relationship between K of the FV and the water level in the tank, h. Third, the inflow into the tank varies based on the FV closure level, which depends on the water level in the tank. The water level in the tank and the corresponding inflows are, in particular, modelled hydraulically using a spreadsheet hydraulic model developed for this purpose. The hydraulic model has specific key inputs: field consumption measurements, K—h relationship, and other hydraulic parameters including: inlet pressure, tank elevation, tank size, and friction and roughness of the service connection between the water meter and the water tank. Using the flow—error relationship of the meter, K—h relationship of the FV, and the hydraulic model, the error of the meter can be computed in the model during the course of the day. Afterward, different sensitivity analyses for different intermittency degrees and tank sizes were conducted by changing the model inputs. Finally, to recognise the effect of the FV type, the above steps were carried out for three different types of FVs. The following parts elaborate on the study's methods and experiments.

8.2.1 Determination of meter errors for extended flow range

Laboratory experiments were conducted at the Bandung Metrology Centre (Indonesia) to determine the errors in customer water meters for extended flow range. Thirteen used meters were collected from the field in Bandung city and replaced with new meters. The meters were class B, multi-jet mechanical meters, with a diameter of ½" (15 mm), representing five types of different manufacturers, and with different ages that ranged between 7 and 19 years. The meters were tested using a standard water meter test bench that includes a water pump, stop valves, flow regulating valves, three flow meters (rotameters, viscosity and gravity type), and a standard water tank, as shown in Figure 8.2. The tests were carried out at pressures up to 4 bar and a temperature of 28 °C. The testing procedures were in accordance with EN 14154-3 and other recommendations (Arregui et al. 2006a; ISO 2014b; OIML 2013b), where the reading of the meters were taken at rest. Different issues were considered during the tests (Arregui et al. 2006a): purge operation tests, flowing direction of the meters, distance between the meters, meter leak checks, eliminating air in the test section, no trapped air pockets, operation of the valve was fast enough to avoid flow uncertainty and slow enough to avoid generating pressure surges, flow rates were adjusted carefully and checked with a stop watch, flow rates did not deviate during the test, and finally the readings of the meters were taken with the highest possible resolution when meters came to a complete stop. The error of each meter was then calculated using Equation 8.1.

$$\varepsilon = \frac{(V_i - V_{i-1}) - V_a}{V_a} \times 100 \qquad (8.1)$$

where ε is the error of the meter (%), V_i is the reading of the meter when the test stopped (m^3), V_{i-1} is the reading of the meter just before the test started (m^3), and V_a is the total volume in the standard tank (m^3).

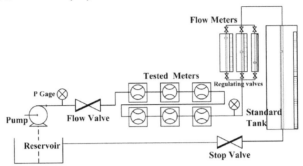

Figure 8.2. Water meter test bench schematics

The meters were tested at five critical flows of the meters: q_{start} (10 l/h), q_{min} (30 l/h), q_t (120 l/h), q_p (1500 l/h), and q_{max} (3000 l/h). However, to enhance the accuracy of the established error curve, the meters were further tested at 10 other flows. These flows were: 15, 20, 35, 40, 200, 300, 400, 500, 600, and 1000 l/h. In total, there were 15 test flow rates, to establish a more detailed and reliable error curve that incorporates all range of expected flows in this study under continuous and intermittent supply conditions.

8.2.2 Investigating the FV characteristics

A laboratory experiment was also carried out at the Bandung Metrology Centre (Indonesia) to assess the hydraulic characteristics of different FV types (mainly the FV resistance coefficient K but also the distance of the FV movement trajectory, commonly known as the modulation range). To simulate the intermittent supply scheme in the lab, additional equipment was required. Three different brands of FVs with two common sizes ½" (15 mm) and ¾" (20 mm) were obtained from the local market and attached to a small water tank with a capacity of 100 l and a cross-sectional area of 2692.3 cm^2. This equipment was connected to a portable ultrasonic flow meter (q_{start} = 6 l/h and q_{min} = 10 l/h) and a pressure regulating valve, as shown in Figure 8.3.

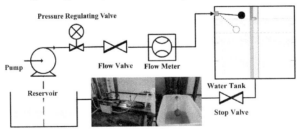

Figure 8.3. Schematics of the FV characteristics analysis experiment

The inflow into the tank is, in principle, the same flow that passes the flow meter. When the water level in the tank rises to the FV level, the FV starts to close, and simultaneously the flow starts to lower. The data obtained from this process is the flow rate and the volume and time of each flow. The ratio of the volume to the cross-sectional area of the tank gives the water level in the tank, h (mm). The difference between h when the FV starts to move and h when the FV is at a complete stop is the modulation range of the FV. Based on the flow data and water level in the tank during different times, the K of the FV can be calculated using Equation 8.2 (Crane_Co. 1957; Mckenzie and Langenhoven 2001).

$$K_i = \frac{P}{Q^2} \qquad (8.2)$$

where K_i is the FV flow resistance coefficient $(m.(l/s)^{-2})$ at a specific water level in the tank h_i (mm), P is the pressure (mwc) before the FV, and Q is the inflow into the tank (l/s). Once the FV characteristics are known, they can be used to simulate the filling process for other hydraulic conditions and tank sizes. The pressure of the experiment was set to 210 kPa when the FV was fully closed. When the FV was fully opened, an inflow of 0.11 l/s occurred with a pressure of 50 kPa. These settings were determined based on preliminary field measurements of flows and pressures at water meters of 30 customers in Bandung city. Field network pressures and flows were measured between November 2018 and January 2019. Measurements were obtained before the water meter using calibrated portable ultrasonic meter. The meter (Linflow PF20A) has a size of 15 mm, resolution of 0.01 l, maximum error of ±1% at 10 l/h, maximum pressure of 10 bar, and maximum flow of 1500 l/h.

The resulting K_i values (58, 52, and 19 K_i values for FV1, FV2, and FV3 respectively) for each corresponding water level in the tank h_i were then studied. A regression analysis was conducted to determine the relationship between the water level in the tank h_i (mm) and the FV coefficient K $(m.(l/s)^{-2})$. To obtain a h—K_i relationship, a non-linear regression analysis was conducted using a specific software tool (CurveExpert Pro 2.6.5). The software library contains several data fitting models including the Bleasdale Model, Hoerl Model, and Logistic Model. The Bleasdale Model presented in Equation 8.3 was selected to model the experimental data.

$$K = (\alpha + \beta h)^{\frac{-1}{\gamma}} \qquad (8.3)$$

where K is the FV resistance coefficient $(m.(l/s)^{-2})$, h is the water level in the tank (mm), and α, β, and γ are the fitting factors of the equation. The values of the α, β, and γ factors were further optimised using the Microsoft Excel Solver, where the total value of the residual sum of squares (RSS) in Equation 8.4 was minimised.

$$RSS = \sum_{i=0}^{n} (K_{experiment} - K_{model})^2 \qquad (8.4)$$

To assess the final fitting of the equation with the optimised α, β, and γ factors, the coefficient of determination R^2 was calculated. As this is a non-liner regression, the key factor to determining the quality of the fit was not R^2 but the minimum value of RSS. Once the factors of Equation 8.3 were optimised, R^2 can be used to indicate the final fitting level. After defining the model that fits the experimental data including its factors, k can be determined at any water level in the tank and for different tank sizes and hydraulic conditions.

8.2.3 Modelling the water level in the tank

The water level in the tank can be modelled using the tank continuity equation (Equation 8.5) based on the inflow and outflow of the tank, as follows:

$$Q_i - Q_{out} = A \frac{dh}{dt} \qquad (8.5)$$

where Q_i is the inflow into the tank, Q_{out} is the discharge out of the tank, and A is the area of the tank. Q_i At any specific time, can be determined based on the Bernoulli and Darcy-Weisbach equations, as shown in Equation 8.6.

$$Q_i = \sqrt{\frac{\frac{P}{\gamma} - (Z_2 - Z_1)}{K_p + K}} \qquad (8.6)$$

where Q_i is the inflow into the tank (l/s), $\frac{P}{\gamma}$ is the network pressure at the water meter (mwc), $Z_2 - Z_1$ is the elevation difference between the inlet of the tank and the water meter (m), K is the FV resistance coefficient, and K_p is the pipe headloss coefficient between the water meter and the inlet of the tank, which can be calculated using Equation 8.7.

$$K_p = \frac{8fL}{\pi^2 g D^5} \times 10^{-6} \qquad (8.7)$$

where K_p is the pipe headloss coefficient (m.(l/s)$^{-2}$) that also considers minor losses, L is the length of the pipe between the water meter and the FV (m), D is the pipe diameter (m), g is the acceleration of gravity (m/s^2), and f is the Darcy-Weisbach friction factor (unit-less) which is calculated by iteratively solving the Colebrook-White equation that is presented in Equation 8.8 (Colebrook and White 1937; Colebrook 1939).

$$\frac{1}{\sqrt{f}} = -2log(\frac{\varepsilon}{3.7\ D} + \frac{2.51}{Re\ \sqrt{f}}) \qquad (8.8)$$

where ε/D is the relative pipe roughness and R_e is the Reynolds number that is calculated using Equation 8.9 (Sommerfeld 1908).

$$R_e = \frac{\rho u D}{\mu} \qquad (8.9)$$

where ρ is the water density (kg/m^3), u is the water velocity (m/s), and μ is the water dynamic viscosity (N.s/m^2).

In contrast, measuring or estimating Q_{out} is crucial in the modelling of the filling and emptying of the tank. The resolution of the Q_{out} data should fit the resolution of the model. In this study, a high-resolution model was built with a time step of 5 s. To incorporate proper data in this model, high-resolution logged measurements were obtained. The measurements of a domestic customer consumption were logged in Castellon, Spain for one week, from March 18 to March 25, 2008. The aim was to model the water level and inflows into the tank using a real high-resolution consumption profile. Any other consumption pattern might amend the calculated weighted error in the model, but will not significantly change the trends of the results. The consumption flows were measured using an oscillating piston meter (Aquadis+ from ITRON). The meter is Class C with size of DN15, q$_{start}$ of 1 l/h, q$_{min}$ of 15 l/h, volume resolution of 0.1 l, and time resolution of 0.02 s. The measurements were logged using Sensus loggers with pulses event times recorded with a resolution of 0.02 s and memory capacity of 256,000 registers. The data were then processed by coding a macro in Microsoft Excel using Visual Basic. The flow data were re-generated every 5 s in the course of the day, to fit the model time step. The instantaneous demand consumption flows during the day were then inserted in the model. Finally, to simulate the filling process in the tank, some parameters were assumed. The elevation difference between the level of the water meter and the tank inlet was assumed to be 3 m, the length of the service connection pipe between the water meter and the tank inlet was 20 m (maximised by 30% to consider other minor losses), water density was 999 kg/m^3, dynamic viscosity was 0.0011 N.s/m^2, roughness of the pipe was 0.1 mm, and finally, a network pressure of 210 kPa at the water meter. The tank size was initially set to a small water tank with a capacity of 500 l and typical dimensions of 1110 mm height, 923.7 mm maximum water level in the tank, and 830 mm diameter. Afterwards, different tank sizes were modelled to analyse the sensitivity of the model to the tank size. Based on all of the above input, the water level in the tank can be modelled at a time step of 5 s and with high-resolution instantaneous consumption.

8.2.4 Analysis of meter inaccuracies

The accuracy of the water meter is a function of the flow rate that passes through the meter. Therefore, the meter accuracy varies according to the different flows that pass through the meter during the day. In fact, the flows that go through the water meter are the consumption flows. However, for intermittent supply with a water tank and FV arrangement, the flows that go through the meter are not the consumption that discharges out of the tank, rather, they are the inflows into the tank. These inflows are influenced by the instantaneous consumption discharges, network pressure, and the water level in the tank. In this study, the error curve of the tested meters was established based on the extended range of 15 test flow rates. The error-flow relation was analysed by non-linear regression analysis using CurveExpert Pro 2.6.5. The best fit model in the tool library was the Rational Model which is presented in Equation 8.10. The factors of the model were optimised, and then the meter error can be determined at any flow.

$$\varepsilon = \frac{a + bQ_i}{1 + cQ_i + dQ_i^2} \tag{8.10}$$

where ε is the error of the meter (%); a, b, c, and d are the equation fitting factors; and Q_i is the inflow rate (l/h). After the flow—error relation is well established, the meter weighted error can be calculated either based on the instantaneous inflows into the tank or by creating a histogram of these flows, and then estimating the meter inaccuracy.

The impact of different flow profiles on meter errors was recognised in this study. A comparison was conducted between the continuous water supply without a water tank and continuous supply with a water tank and FV. This was done by calculating the meter errors for the histogram of the consumption flows (discharges out of the tank) and for another histogram of the inflows into the tank. The results were also compared to the meter errors that were generated considering two other typical flow profiles in the literature (Arregui et al. 2015; Arregui et al. 2006a).

8.2.5 Investigation of the impact of intermittency

The impact of intermittency on the meter performance was investigated by imposing different intermittency levels in the model and then analysing the inflows and the water meter errors. Some systems are intermittent supply with one or two full supply days during the week. In this case, the supply becomes continuous supply for one or two days. Other assigned supply scenarios included very short supply time: 2 h/d; short supply time: 4 h/d; average supply time: 8 h/d; long supply time: 20 h/d; and continuous supply: 24 h/d. These intermittency levels are similar to realistic intermittent supplies (IBNET 2020). The 20 h/d supply scenario was assumed to occur from 00:00 AM to 12:00 PM and then from 16:00 PM to 00:00 AM. Thus, the intermittency in this scenario occurred for four hours between 12:00 PM and 16:00 PM. The inflows into the tanks were studied in this scenario and the meter error was calculated at a time step of 5 s. This approach was implemented for the supply scenario of 8 h occurring between 8:00 AM and 16:00 PM, the supply scenario of 4 h/d occurring between 8:00 AM and 12:00 PM, and the supply scenario of 2 h/d occurring between 10:00 AM and 12:00 PM. After the different intermittency scenarios were analysed, the impact of the intermittency on the meter performance can be concluded and compared to the situation of continuous water supply.

8.2.6 Sensitivity analysis - different tank sizes

Finally, the impact of various water tank sizes with FV on the performance of the water meter was assessed. This was conducted by simply altering the tank capacity in the model and analysing the resulting inflows into the tank.

8.3 RESULTS AND DISCUSSION

8.3.1 Meter error curves

Figure 8.4 shows the error curves of the tested meters: Figure 8.4a shows the individual error curves for each meter, Figure 8.4b shows the average error curve for each meter type where the same type of meters were grouped, and Figure 8.4c shows the average error curve of all the tested meters. The meter performance clearly varies, with high error at low flows and a better accuracy at higher flows. At a certain threshold in the lower zone, the meter starts to slightly over-register the passing flow, and this becomes the trend for the permanent flow in the upper zone. Figures 8.4a and 8.4b show that the performance of the water meter varies because each individual meter differs in terms of age, type, and operating conditions. The tested meters were, therefore, grouped based on their age, size, and type, to minimise the variance and propose a sound meter replacement policy. These groups of the tested meters based on their age and model are presented in the Supplementary Material 8.5.1.

Figure 8.4c shows the average errors for q_{start}, q_{min}, q_t, q_p, and q_{max}. For these critical flows, the average errors and the standard deviations, shown in brackets, are -85.06% (19.74%), -11.7% (30.27%), 2.21% (2.75%), 0.96% (2.61%), and 1.99% (3.54%), respectively. The average curve error in Figure 8.4c was used in the subsequent analyses because the focus of this research was not to propose a meter management policy but to figure out the impact of the FV on the meter performance.

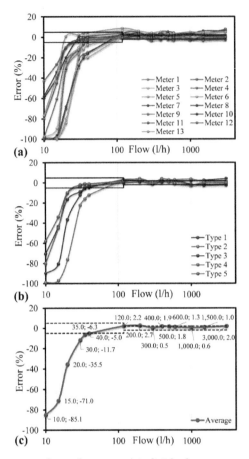

Figure 8.4. Error curves of tested meters: a) individual error curves, b) error curves by meter type, and c) average error curve

8.3.2 FV characteristics

The detailed characteristics of the FVs are presented in the Supplementary Material 8.5.2. The valves have different sizes and modulation ranges. The modulation ranges are 60.74 mm, 38.50 mm, and 18.19 mm and the sizes are ½" (15 mm), ¾" (20 mm), and ¾" (20 mm) respectively for FV1, FV2, and FV3. However, as the nozzle and washer of the FVs can differ in size and discharge, the hydraulic characteristics of the FVs can be expressed in the form of the K values of the FV. Figure 8.5 shows a plot of the K values of the three FVs as a function of the valve openings. The non-linear regression model that gave the best fit for this data was the Bleasdale Model (Equation 8.3), with R^2 of 0.97, 0.94, and

0.99 for FV1, FV2, and FV3, respectively. R^2 is, in fact, not sufficient to indicate the fitting of the experimental data. Figure 8.5 shows the difference between the modelled and experimental K values of the FVs. Deviations occurred between the modelled and experimental K values toward the fully open status of FV2, and to a lesser extent, of FV1. The impact of the K deviations on the hydraulic model is not critical. The K equations were substituted in the model with other exponential equations to test the sensitivity of the model to the K equations, and the weighted errors generated by the model were not affected significantly. For the purpose of this study, the Bleasdale model can be used to determine the K value of each FV at any height of the tank, which is important in determining the inflow into the tank as shown in Equation 8.6. The α, β, and γ fitting parameters of the Bleasdale Model are also presented in Figure 8.5.

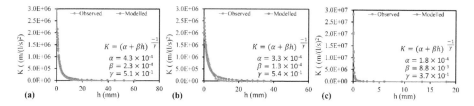

Figure 8.5. Experimental and modelled FV coefficients at different openings of the valves: (a) FV1, (b) FV2, and (c) FV3

8.3.3 Impact of different consumption patterns

Figure 8.6 shows different consumption patterns of domestic water usage. The water volume (l) for each flow range (l/h) was summed up and divided by the total consumption volume. Figure 8.6a shows a typical consumption pattern for urban households (Arregui et al. 2006a), Figure 8.6b presents an example of a consumption pattern for a household with a water tank (Arregui et al. 2015), and Figure 8.6c presents a histogram of the field consumption measurements in Spain that were obtained for this study. The meter performance varies when the flow that passes through the meter varies. For this reason, the weighted error according to the consumption is crucial in indicating the performance of the meter in the field (Arregui et al. 2006a; Shields et al. 2012). Based on the used meter experiments and the established average error curve, the weighted errors of FV1 were -4.2%, -8.7%, and -0.7% for consumption patterns a, b, and c, respectively. The consumption pattern in Figure 8.6c has a significantly lower proportion of the lowest flow range, and thus a lower error level.

151

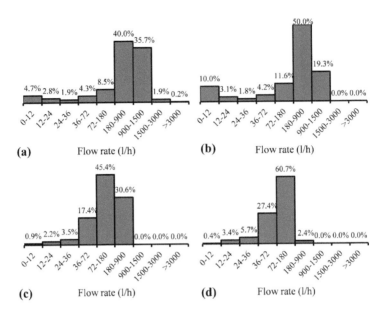

Figure 8.6. Histogram of four consumption patterns: (a) typical consumption profile (Arregui et al. 2006a), (b) typical consumption profile for premises with water tanks (Arregui et al. 2015), (c) consumption pattern based on field measurements in Castellon,Spain, and (d) histogram of modelled inflows into the tank corresponding to consumption outflows of the tank

Calculating several metering errors for the same meter sample confirms the significance of the weighted error methodology in indicating the performance of the meter. Furthermore, the consumption pattern in Figure 8.6c represents the consumption pattern after the tank, which is used by the customer. However, once this consumption pattern is modelled instantaneously for the water tank and FV model, the critical flows will not be the consumption flows, but the corresponding inflows into the tank (the flows that pass through the meter). Interestingly, these tank inflows are generally lower than the consumption flows. This could be the reason to report meter accuracy at the inlet of the tank lower than the meter accuracy at the outlet of the tank (Criminisi et al. 2009; Rizzo and Cilia 2005). The low flow rates in Figure 8.6d are more significant as in the flow rate ranges of 12-24 l/h, 24-36 l/h, and 36-72 l/h. Therefore, the flow pattern in Figure 8.6d has a higher weighted meter error (-1.24%), as elaborated further in the following section.

8.3.4 Modelling the water level, FV, meter error, and continuous supply

The left panels of Figure 8.7 show the modelled water level in the tank, while the right panels show the modelled inflow into the tank with its corresponding meter error for each

FV. The instantaneous consumption lowers the water level in the tank, and immediately thereafter, the FV opens and a refill process begins. The refilling process differs between the FVs depending on the hydraulic characteristics of the FVs (K). The difference between the inflow and the outflow of the tank causes the water level in the tank to fluctuate. Interestingly, Fig. 4 also confirms that the inflows in the tank are constantly less than the consumption discharges out of the tank. Thus, for the same consumption pattern, the water tank and FV reduce the flows that pass the water meter, resulting in higher meter errors.

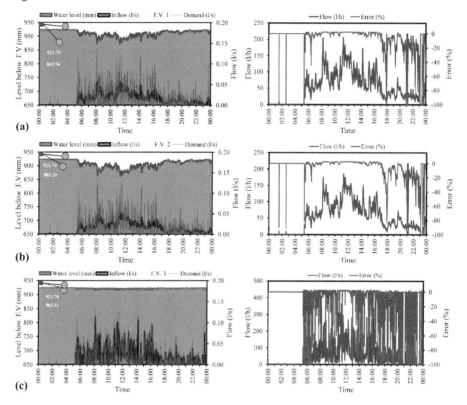

Figure 8.7. Water level in the tank (left) and meter error corresponding to the inflow (right) for different types of FVs: (a) FV1, (b) FV2, and (c) FV3. (Continuous supply, 24 h/d)

Different FVs have different hydraulic characteristics which influence the inflow rates and water level in the tank, ultimately influencing the water meter performance. Figure 4.7a shows the water level and meter error with FV1 (size ½" [15 mm], modulation range 60.74 mm) in a typical tank with a net height of 923.7 mm. The error of the water meter oscillates in accordance with the inflow into the tank, yet with an average error of -7.47%

153

(for 76% of the time, when there is an inflow into the tank) and a weighted error of - 1.24%. Figure 4.7b shows the water level and meter error with FV2 (size ¾" [20 mm], modulation range 38.50 mm); the average error was -6.88% and weighted error was - 1.12%. Figure 4.7c shows the water level and meter error with FV3 (size ¾" [20 mm], modulation range 18.19 mm); the average error at -9.77% and weighted error at -1.74%. Although FV1 and FV2 are different in size and modulation range within the tank, they have close hydraulic characteristics and error levels. This is because the size and modulation range of the FV, in addition to the sizes of the nozzle and washer inside the FV, influence the discharge rate of the FV, as illustrated in the Supplementary Material 8.5.2.

Conversely, FV2 and FV3 are the same size. However, FV3 allows higher inflows, refills the tank faster, and introduces higher flows. As a result, the water level remains close to the full level in the tank. Typically, the water meter performance improves with higher flow rates and worsens at low flow rates. Nonetheless, FV3 has higher flows than FV1 and FV2, but interestingly FV3 has higher errors, as shown in Figure 4.7c. To explain this, when an instantaneous consumption is imposed, for every instantaneous consumption a flow immediately comes in and fill the tank to its full level. Thus, the inflow into the tank grows from zero to a certain value and then falls back to zero. In the case of FV3, this process is fast and the cycle is completed before the next instantaneous consumption occurs. On the contrary, the discharge rate of FV1 and FV2 is lower, resulting in lower inflow rates and a slower filling process. The inflow to the tank grows from zero to a certain level and then falls back, but the next instantaneous consumption occurs before the inflow returns to zero, increasing again the inflows and causing further drops in the water level in the tank. This process, in the case of FV1 and FV2, causes the ultimate low flows to occur less frequently during the day. As a result, the low flows of FV1 and FV2 occurring during the day are larger in value and for shorter durations than the low flows of FV3, and therefore, the error level of FV3 is higher than the error levels of FV1 and FV2.

To conclude, for the case of combining continuous water supply with an arrangement of water tank and FV, the performance of the water meter improves when the FV discharge rate is lower. This implies that the use of smaller FVs results in better water meter performance and bigger FVs accommodate larger flows but cause higher meter errors. The influence of the length of the modulation range of the FV on the discharge rate of the FV should be investigated. If FVs with a shorter modulation range have higher discharge rates, then a longer modulation range of the FV should trigger better meter performance.

8.3.5 Impact of intermittency on the meter performance

Figure 8.8 presents the water level in the tank and meter error within a supply time of 8 h/d with three different FVs. In this scenario, water is supplied between 8:00 AM and

16:00 PM. Although the supply time is only 8 h/d, the timing of the supply corresponds to the consumption pattern that discharges out of the tank, such that the water level in the tank remains above zero. Thus, there is always water for consumption during the course of the day. Other supply time scenarios have been analysed, too. The Supplementary Material 8.5.3 shows the meter error and water level in the tank with a supply time of 2 h/d, 4 h/d, and 20 h/d, where the water level in the tank drops to zero and the consumption in this case was loaded uniformly during the following supply time. As shown in Figure 8.8, when the water level drops in the tank, the FV opens fully and higher inflows occur. The higher the intermittency level, the longer the FV remains open, and the higher will be the inflows into the tank.

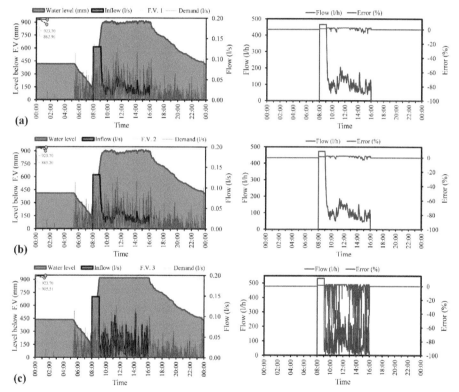

Figure 8.8. Water level in the tank (left) and meter error (right) for a supply of 8 h/d for different FVs: (a) FV1, (b) FV2, and (c) FV3. (Intermittent supply, 8 h/d)

Expectedly, the meter performance improves in intermittent supply, and the lower the number of supply hours, the lower the under-registration level of the meter. Moreover, when the system regime is intermittent supply, the entrapped air in the network discharges through the water meters, causing further over-registration of the mechanical meters. In this case, longer intermittencies result in more entrapped air in the network, and thus larger volumes of air will be measured as water.

Table 8.1 shows the water meter errors under different intermittency levels, for the three FVs, and without considering the over-registration caused by the air pockets travelling through the meters. The trend is nearly consistent; less under-registration was observed when more intermittency was imposed. At a certain degree of intermittency, the average error of the water meter becomes higher than zero, and the meter error is on average not under-registered but over-registered. Even though this is profitable for the water utility, it is important to consider that supply intermittency has other severe consequences on the network, e.g. meter damage, more transients, higher burst rates, more network fatigue, water quality deterioration, and poorer service level.

Table 8.1. Meter error with different FVs for different intermittency levels

Supply	Tank size	FV1		FV2		FV3	
		Avg. error	Weig. error	Avg. error	Weig. error	Avg. error	Weig. error
2 h/d	500 l	1.3	1.3	1.3	1.3	1.3	1.2
4 h/d	500 l	2.0	1.7	2.0	1.8	1.4	1.5
8 h/d	500 l	1.3	1.6	1.5	1.6	-4.5	0.7
20 h/d	500 l	-4.3	0.7	-3.8	0.8	-2.6	0.8
24 h/d	500 l	-7.6	-1.2	-7.0	-1.1	-9.8	-1.7

8.3.6 Impact of the tank size on the meter performance

Figure 8.9 shows the water level, consumption, and inflow into the tank with FV1 and four different tank sizes: 500 l, 1000 l, 2000 l, and 5000 l. This analysis is conducted for the continuous supply scenario. These tank sizes have the same height (923.7 mm), and thus the tank size is changed by adjusting the surface area of the tank through increasing the diameter of the tank. When the tank size increases, the consumption event causes a lower decrease in the water level in the tank, and thus lower FV inflow into the tank. The instantaneous consumption is the same for the four tanks, yet the inflow and the water level in the tank is smoother and with less spikes as the tank size becomes larger. This is because the balancing effect of the tank is more significant when the tank size is bigger. The instantaneous consumption has less impact on the water level, and therefore, the water level falls slightly in the tank, and the water level curve in Figure 8.9 becomes more uniform. When the inflows into the tank are more uniform and with less spikes, the range of flows shrinks and the ultimately low flows occur less during the day.

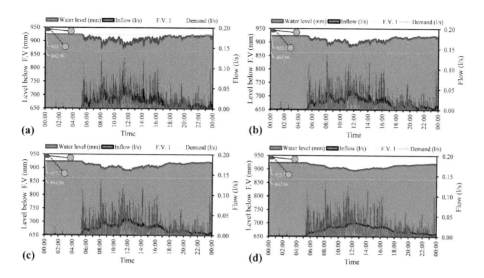

Figure 8.9. Water level in the water tank with FV1 for different tank sizes: (a) 500 l, (b) 1000 l, (c) 2000 l, and (d) 5000 l. (Continuous supply)

Figure 8.10 shows the inflow and the corresponding meter errors with FV1 and different tank sizes. The inflow values in l/s for each time step of 5 s are very low, and therefore converted to l/h. The Supplementary Material 8.5.4 shows the water level, range of inflows, and meter errors for the tank arrangements with FV2, FV3 and different tank sizes. The trend remains the same. The larger the tank size, the more time is required to fill the tank back to its full level, and the more uniform its inflows become. With FV1, the average errors of the meter were -7.64%, -7.46%, -6.60%, and -4.75% with tank sizes of 500 l, 1000 l, 2000 l, and 5000 l, respectively. Similarly, the weighted errors with FV1 were -1.24%, -1.33%, -1.26%, and -1.03% with tank sizes of 500 l, 1000 l, 2000 l, and 5000 l, respectively. The Supplementary Material 8.5.4 presents the same results for the tank arrangements with FV2 and FV3.

Unexpectedly, these results suggest that larger tank sizes confer better meter performance and less error levels because the inflows into the tank is more uniform. The governing factor in this specific problem is not the size of the tank, but the time required to refill the tank and bring the water level in the tank back to its full level. With larger tank sizes, the water level drops slightly in the tank with consumption and the FV opens slightly, causing a lower inflow rate that requires a longer time to refill the tank. If this 'filling time' is shorter than the time difference between the first and the next instantaneous consumption, then higher metering errors will occur with bigger tank sizes. For example, if the water customer uses water every 30 s and the tank refilling takes 15 s, then the tank inflow grows to a specific value and returns toward zero as the FV closes. The inflow is zero when the refilling process is complete. In this case, larger tank sizes introduce more and longer low inflows and eventually cause more metering errors. However, if the refilling process takes 40 s and the consumption events occur every 30 s, then the refilling process starts after the occurrence of consumption. The inflow value increases from zero to a certain value and then decreases towards zero. After 30 s and before the inflow reaches zero, another consumption occurs, increasing the inflow value again. Consequently, there are less ultimately low flows (close to zero) and they occur for shorter times during the course of the day. In this case, larger tank sizes cause relatively lower metering errors.

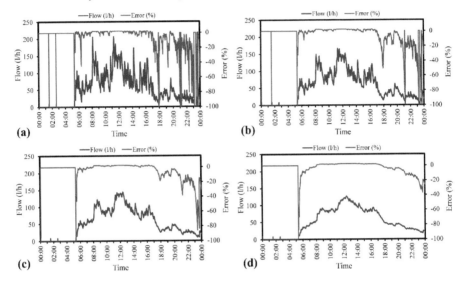

Figure 8.10. Meter error with FV1 for different tank sizes: (a) 500 l, (b) 1000 l, (c) 2000 l, and (d) 5000 l. (Continuous supply)

To clarify this point further, Figure 8.11 shows the inflow range that discharges into the tank for two tank sizes with FV3 during two hours in the day. The frequency and depth of encroachment into the poor performance zone is greater for the smaller tank size than for the larger tank size. This phenomenon is the reason for the better meter performance of the bigger tank size. The Supplementary Material 8.5.4 represents the full spectrum of inflows for the different arrangements of FVs and tank sizes. Figure 8.11 and The Supplementary Material 8.5.4 suggest the same conclusion; the maximum inflow is lower when the tank size is larger, and the minimum inflow is greater because the time required to reach the full tank level is longer with larger tank sizes. In this situation, the next instantaneous consumption occurs before the tank is completely full, causing the inflow to rise again. This process influences the range of inflows into the tank which, in turn, influences the meter performance. In this case, the meter performance improves when the tank size is larger.

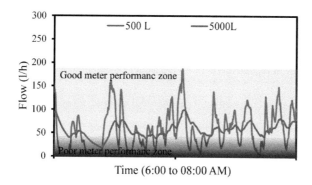

Figure 8.11. Inflows with FV3 for two different tank sizes during two hours in the day. (Continuous supply)

To summarise this discussion, the impact of the tank size on the meter performance depends on the time required to refill the tank and the time difference between the consumption events during the day. While the consumption concentration depends on the consumption pattern of the water users, the refilling time is influenced by K, the network flow and pressure, and the water level in the tank. If the refilling time is longer than the time difference between consumption events, then bigger tank sizes will allow for less low flows, and thus less errors and apparent losses occur, and vice versa.

Finally, the shape of the tank is also important. A tank with the same size can have different surface areas. For example, a tank of 2 m³ capacity can be with 1 meter in height and 2 m² in surface area or 1 m² in surface area and 2 m in height. The latter shape is vividly better in terms of meter performance because the surface area is smaller, and thus the consumption event causes higher drops in the water level inside the tank and higher inflow rates. Hence, a cylindrical tank installed in a horizontal position will result in fewer errors than a vertical position or a square tank.

8.4 CONCLUSIONS

This study analysed the impact of the water tank and FV arrangement on the performance of water meter under intermittent and continuous water supply conditions. The error curve of a sample of used water meters was established, the FV characteristics were experimentally determined, and the water level in a water tank with a FV was hydraulically modelled. Typically, water meter accuracy declines as the flow rate that passes through the meter decreases, and decays rapidly when the flow rate is less than the minimum flow of the meter. The water consumption pattern of a user is, therefore, a critical factor influencing the meter performance. The proportion of water consumption at low flows (0-36 l/h) out of the total consumption substantially affects the meter accuracy, confirming the significance of the weighted error as an indicator of the actual field performance of a water meter. Considering the sensitivity of the low flows, the effect of the FV and water tank arrangement on the meter performance should be recognised.

8.4.1 Effect of the combination of FV, water tank, and continuous supply

Once the arrangement of water tank and FV is in place downstream of the water meter, the flows that pass the water meter are not the consumption flows that discharge out of the water tank, but the inflows into the tank. The inflows into the tank are consistently lower than the outflows due to the balancing nature of the tank. This result suggests that the arrangement of water tank and FV, when combined with continuous supply, worsens the water meter performance and causes higher metering errors and higher apparent losses. In addition, different types of FVs have different hydraulic characteristics which influence the range of the inflow rates into the tank, the water level in the tank, and ultimately influence the water meter performance. This effect applies in continuous supply. The performance of the water meter improves once the FV discharge rate is lower. With higher FV discharge rates, the water level remains close to the full level in the tank causing only slight FV opening and generating more ultimately low inflows during the day. This implies that smaller FV sizes are advantageous to improve water meter performance and bigger FV sizes cause more metering errors.

For a continuous supply system combined with a water tank and FV arrangement, the sensitivity of the meter accuracy to the size of the water tank is variable. The critical factor is the time required to refill the tank and the time between the instantaneous consumption events. If the refilling process of the tank is complete before the next consumption starts, then, in this case, bigger tanks cause longer period of low flows and produce larger metering errors. However, if the speed of the refilling process is not sufficiently fast and next consumption event occurs before the tank is completely full, then, in this case, bigger tank trigger larger inflow rates and results in relatively lower metering errors and less apparent losses. Furthermore, the shape of the tank is also important. A cylindrical tank installed in a horizontal position will result in fewer errors than a cylindrical tank in vertical position or a square tank. If the tank is almost full, the water surface area at the beginning of the filling process is very small, causing the water level to drop quickly and increasing the FV inflows.

8.4.2 Effect of degree of intermittency

The impact of supply intermittency on the water meter performance is central. The FV in intermittent supply drops down in the water tank and does not remain within the modulation range of the FV, as in the case of continuous supply. Four different intermittency levels were assessed: 2 h/d, 4 h/d, 8 h/d, and 20 h/d. The results are consistent across these levels. Higher intermittency levels improve the meter performance and reduce metering under-registration. This is because the higher the intermittency level, the longer the FV remains completely open, and the higher will be the inflows into the tank. With high intermittencies, the average error of the water meter becomes greater than zero, which means that the meter error is on average not under-registration but over-registration. Therefore, it is important for water utilities to consider in their meter replacement policy higher metering errors and more apparent losses when transforming from intermittent to continuous supply, to make the transformation to 24/7 system feasible and efficient.

8.5 SUPPLEMENTARY DATA

The following section presents additional information on meter error analysis, FV characteristics, meter performance at different intermittency levels, and meter performance at different tank sizes.

8.5.1 Meters' errors and grouping

Table 8.2. Individual meters type, age and errors (%) at different test flows (l/h)

#	Type	Age	10	15	20	30	35	40	120	200	300	400	500	600	1000	1500	3000
1	Type 1	10.0	-97.6	-98.4	-32.0	-14.8	-6.4	-5.4	1.6	2.4	1.9	1.3	-2.4	-1.1	-1.0	1.3	7.9
2		12.0	-98.0	-99.0	-31.0	-10.0	0.8	1.0	3.7	1.5	-2.1	1.0	-1.7	-3.2	2.7	3.4	7.4
3		13.0	-97.6	-89.2	-24.4	-10.4	0.8	1.9	6.3	5.3	1.2	3.6	2.4	1.9	2.2	1.5	2.3
4		9.0	-98.8	-98.4	-70.0	-21.6	-16.8	-14.8	1.4	2.2	0.9	-2.6	1.6	2.3	3.3	-1.1	-1.3
5		19.0	-100.0	-100.0	-80.0	-30.0	-20.0	-20.0	0.0	1.7	-3.3	-1.3	2.2	2.6	-1.7	2.0	2.0
6		8.0	-60.0	-40.0	-20.0	0.0	0.0	0.0	0.0	1.5	3.6	2.2	4.2	1.4	3.2	6.0	6.0
7		7.0	-76.8	-38.0	0.8	2.4	4.0	4.4	8.8	6.2	4.3	5.3	2.3	3.5	5.1	4.3	4.6
8	Type 2	13.0	-98.0	-98.0	-84.0	-17.0	-12.0	-8.0	0.5	1.2	2.0	2.8	0.2	-2.3	-1.6	-0.8	0.0
9		7.0	-100.0	-96.0	-68.0	-28.0	-20.0	-16.0	3.5	4.6	-2.3	5.3	5.3	4.3	-2.0	0.2	0.6
10	Type 3	8.0	-98.0	-60.5	-23.0	-13.0	-6.6	-3.2	0.2	2.7	-2.8	3.3	2.5	3.5	-1.1	-2.1	-2.4
11		9.0	-43.0	-26.5	-10.0	-4.5	-3.0	-3.0	-0.8	-1.2	4.2	-2.2	2.5	1.6	3.6	-3.3	-2.6
12	Type 4	17.0	-58.0	-34.0	-10.0	-3.5	-1.0	-1.0	2.0	4.1	0.6	2.3	1.3	2.9	-2.6	1.4	1.6
13	Type 5	7.0	-80.0	-45.0	-10.0	-2.0	-1.0	-1.0	1.5	2.4	-1.4	3.6	2.9	-1.0	-2.6	-0.2	-0.4
Average			-85.1	-71.0	-35.5	-11.7	-6.3	-5.0	2.2	2.7	0.5	1.9	1.8	1.3	0.6	1.0	2.0
	Max		-43.0	-26.5	0.8	2.4	4.0	4.4	8.8	6.2	4.3	5.3	5.3	4.3	5.1	6.0	7.9
	Min		-100.0	-100.0	-84.0	-30.0	-20.0	-20.0	-0.8	-1.2	-3.3	-2.6	-2.4	-3.2	-2.6	-3.3	-2.6
	STDev		19.7	30.3	29.4	10.4	8.3	7.6	2.7	1.9	2.7	2.6	2.1	2.4	2.7	2.6	3.5

Table 8.3. Grouping the average error (%) by model and age

Brand	Age	10	15	20	30	35	40	120	200	300	400	500	600	1000	1500	3000
Type 1	5-10 (yr)	-97.6	-98.4	-32.0	-14.8	-6.4	-5.4	1.6	2.4	1.9	1.3	-2.4	-1.1	-1.0	1.3	7.9
		-98.8	-98.4	-70.0	-21.6	-16.8	-14.8	1.4	2.2	0.9	-2.6	1.6	2.3	3.3	-1.1	-1.3
		-60.0	-40.0	-20.0	0.0	0.0	0.0	0.0	1.5	3.6	2.2	4.2	1.4	3.2	6.0	6.0
		-76.8	-38.0	0.8	2.4	4.0	4.4	8.8	6.2	4.3	5.3	2.3	3.5	5.1	4.3	4.6
	avg.	-83.3	-68.7	-30.3	-8.5	-4.8	-3.9	2.9	3.1	2.7	1.5	1.4	1.5	2.6	2.6	4.3
	11-15	-98.0	-99.0	-31.0	-10.0	0.0	1.0	3.7	1.5	-2.1	1.0	-1.7	-3.2	2.7	3.4	7.4
		-97.6	-89.2	-24.4	-10.4	0.8	1.9	6.3	5.3	1.2	3.6	2.4	1.9	2.2	1.5	2.3
	avg.	-97.8	-94.1	-27.7	-10.2	0.4	1.4	5.0	3.4	-0.5	2.3	0.4	-0.6	2.4	1.5	4.9
	16-20	-100.0	-100.0	-80.0	-30.0	-20.0	-20.0	0.0	1.7	-3.3	-1.3	2.2	2.6	-1.7	2.0	2.0
Type 2	5-10	-100.0	-96.0	-68.0	-28.0	-20.0	-16.0	3.5	4.6	-2.3	5.3	5.3	4.3	-2.0	0.2	0.6
	11-15	-98.0	-98.0	-84.0	-17.0	-12.0	-8.0	0.5	1.2	2.0	2.8	0.2	-2.3	-1.6	-0.8	0.0
Type 3	5-10	-98.0	-60.5	-23.0	-13.0	-6.6	-3.2	0.2	2.7	-2.8	3.3	2.5	3.5	-1.1	-2.1	-2.4
		-43.0	-26.5	-10.0	-4.5	-3.0	-3.0	-0.8	-1.2	4.2	-2.2	2.5	1.6	3.6	-3.3	-2.6
	avg.	-70.5	-43.5	-16.5	-8.8	-4.8	-3.1	-0.3	0.8	0.7	0.5	2.5	2.6	2.6	-2.7	-2.5
Type 4	16-20	-58.0	-34.0	-10.0	-3.5	-1.0	-1.0	2.0	4.1	0.6	2.3	1.3	2.9	3.6	1.4	1.6
Type 5	5-10	-80.0	-45.0	-10.0	-2.0	-1.0	-1.0	1.5	2.4	-1.4	3.6	2.9	-1.0	-2.6	-0.2	-0.4

8.5.2 Float-valve characteristics

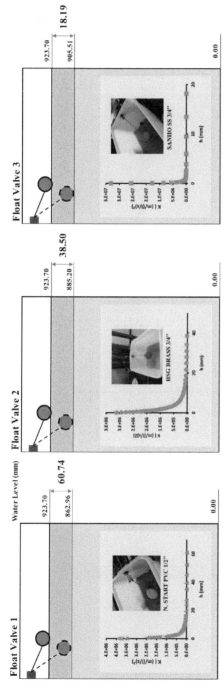

Figure 8.12. Size, modulation range and K of the FVs

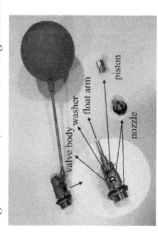

Figure 8.13. Components of the FV

8.5.3 Water level and meter error for different intermittency levels

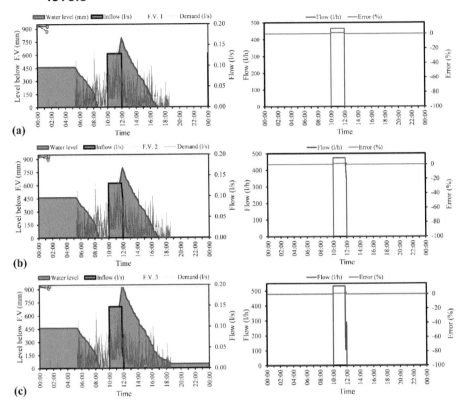

Figure 8.14. Modelled water level within the tank and the meter error for a supply time of 2 hours per day from 10:00 AM to 12:00 AM: (a) FV1; (b) FV2; and (C) FV3

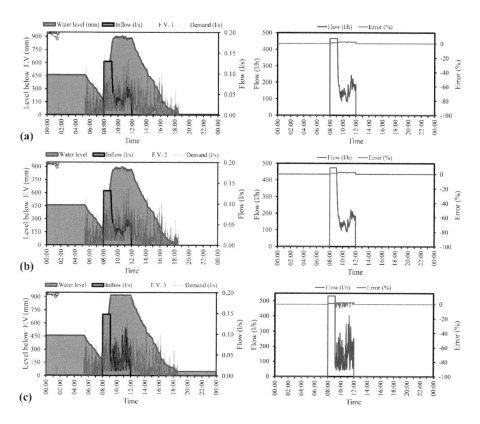

Figure 8.15. Modelled water level within the tank and the meter error for a supply time of 4 hours per day from 08:00 AM to 12:00 AM: (a) FV1; (b) FV2; and (C) FV3

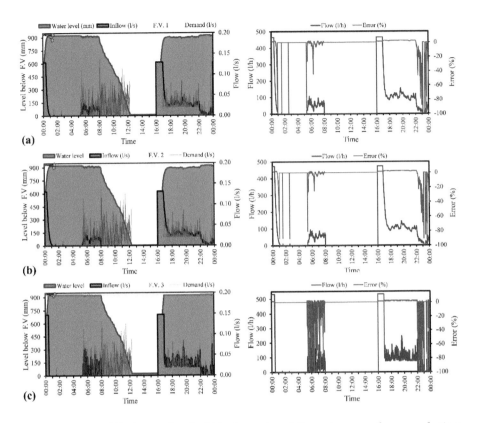

Figure 8.16. Modelled water level within the tank and the meter error for a supply time of 20 hours per day with intermittency between 12:00 AM and 16:00 PM: (a) FV1; (b) FV2; and (C) FV3

8.5.4 Water level and meter error for different tank sizes

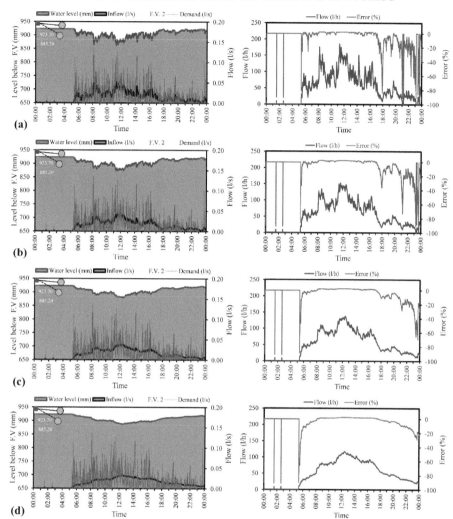

Figure 8.17. Water level and meter error for FV 2 with different tank sizes and for continuous supply scenario (worst case scenario), (a) tank size 500 l, (b) 1000 l, (c) 2000 l, and (d) 5000 l

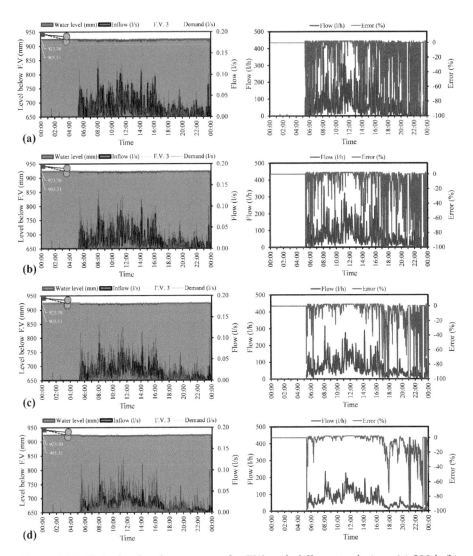

Figure 8.18. Water level and meter error for FV3 with different tank sizes, (a) 500 l, (b) 1000 l, (c) 2000 l, and (d) 5000 l

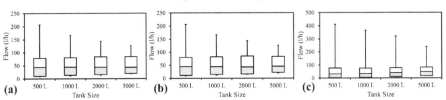

Figure 8.19. Range of inflows into the tank for different tank sizes excluding zero flow, (a) FV 1, (b) FV 2, and (c) FV3

168

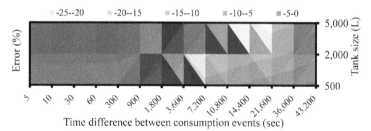

Figure 8.20. Impact of consumption event time on the meter performance

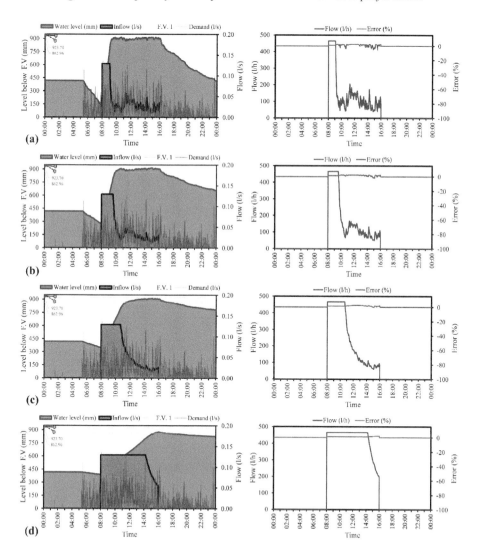

Figure 8.21. Water level and meter error for FV 1 with different tank sizes and under intermittent supply condition (8 h/d), (a) tank size 500 l, (b) 1000 l, (c) 2000 l, and (d) 5000 l

Table 8.4. Summary table of meter error (%) for different FVs, different tank sizes, different intermittency degree, and half-full tank at the start of the simulation: weighted errors (left), and average errors (right)

Supply	Tank Size	FV1	FV2	FV3	Supply	Tank Size	FV1	FV2	FV3
2 h/d	500 l	1.31	1.31	1.22	2 h/d	500 l	1.31	1.31	1.28
	1000 l	1.31	1.30	1.18		1000 l	1.31	1.29	1.18
	2000 l	1.31	1.3	1.18		2000 l	1.31	1.29	1.18
	5000 l	1.31	1.3	1.18		5000 l	1.31	1.29	1.18
4 h/d	500 l	1.74	1.75	1.48	4 h/d	500 l	1.98	2.01	1.42
	1000 l	1.62	1.64	1.42		1000 l	1.85	1.88	1.54
	2000 l	1.42	1.45	1.30		2000 l	1.48	1.52	1.47
	5000 l	1.31	1.3	1.18		5000 l	1.31	1.3	1.18
8 h/d	500 l	1.55	1.61	0.69	8 h/d	500 l	1.29	1.48	-4.51
	1000 l	1.60	1.64	0.99		1000 l	1.62	1.72	-2.28
	2000 l	1.53	1.57	1.20		2000 l	1.63	1.72	0.13
	5000 l	1.38	1.40	1.19		5000 l	1.43	1.45	1.09
20 h/d	500 l	0.72	0.77	0.77	20 h/d	500 l	-4.27	-3.75	-2.63
	1000 l	0.76	0.78	0.70		1000 l	-4.21	-4.21	-3.56
	2000 l	0.85	0.91	0.72		2000 l	-3.90	-3.27	-3.60
	5000 l	1.60	1.70	0.87		5000 l	1.53	1.81	-3.24
24 h/d	500 l	-1.24	-1.12	-1.74	24 h/d	500 l	-7.64	-6.96	-9.77
	1000 l	-1.33	-1.16	-1.88		1000 l	-7.46	-6.59	-9.41
	2000 l	-1.26	-1.11	-1.85		2000 l	-6.60	-5.80	-8.92
	5000 l	-1.03	-0.87	-1.56		5000 l	-4.75	-4.15	-7.48

9

ASSESSMENT OF UNAUTHORISED CONSUMPTION

Estimating unauthorised consumption (UC) in water distribution networks is a challenging task due to its complex and hidden nature. Assigning default values for UC is therefore the common approach. The assumptions made in estimating UC significantly affect the water balance establishment and subsequent planning of water loss reduction measures, which is more critical when UC is prevalent, as is the case in many low- and middle-income countries. This study investigates five possible methods to estimate UC, introducing the background concepts of the methods and demonstrating their applications in six case studies in developing countries. Assuming UC at common default values shows the same UC level over time and does not reflect UC management. UC estimation based on minimum night flow data for a portion of a network is unreliable because UC level varies across the network based on the socio-economic conditions. The water and wastewater balance method objectively estimates the UC volume. However, a prerequisite of this method is the existence of central sewers for part or all of the water network. Analysing the different UC components is theoretically possible. However, this method is susceptible to underestimating the UC because water utilities cannot recognise or detect all illegal uses in the network. Nevertheless, this analysis enables water utilities to understand the nature and types of UC in the network. When an objective and reliable UC estimation is not possible through the above methods, this chapter proposes a matrix for initial UC estimates that uses the number of disconnected customers as an indicator of UC in the network.

This chapter is based on a paper to be submitted to Water and Environment Journal: Al-Washali, T., Sharma, S., Mphahlele, T., and Kennedy, M. "Methods of Assessment of Unauthorised Consumption in Water Distribution Networks".

9.1 INTRODUCTION

Apparent losses include customer meter inaccuracies, consumption data acquisition errors, unmetered consumption underestimations, and unauthorised consumption (UC) (Lambert and Hirner 2000; Vermersch et al. 2016). Managing apparent losses requires implementing specific techniques to reduce these four components as elaborated in Chapter 1. The reduction of UC involves reducing (i) illegal connections by consumers that are not registered in the water utility database; (ii) meter tempering and bypasses established by registered customers; and (iii) water theft from water distribution network equipment such as hydrants and discharge valves (AWWA 2016; Carteado and Vermersch 2010; Thornton et al. 2008; UNHSP 2012). A reduction in UC is attainable through the detection and field inspection of the above UC components (AWWA 2016; Thornton et al. 2008; UNHSP 2012). Customer management; community participation, awareness, and communication policies; and customer surveys are essential in controlling UC (Carteado and Vermersch 2010; Farley et al. 2008). Estimating UC in water distribution networks is a challenging task that significantly affects leakage estimations (as elaborated in Chapters 2 and 5) and leakage reduction planning (as discussed in Chapters 6 and 7). This study investigates possible methods for estimating UC in water distribution networks by first conceptualising the methods and then applying them to six water distribution networks in developing countries. The research analyses the impacts and implications of UC estimations on water loss management and provides recommendations for a practical UC estimation approach in low- and middle-income countries (LMIC). An improved UC estimation can substantially contribute to UC minimisation, accurate leakage estimation, more realistic leakage reduction planning, and more effective water loss management in water distribution networks.

9.2 UNAUTHORISED CONSUMPTION ASSESSMENT METHODS

The most common approach for water balance establishment is the top-down water audit, where the apparent losses are estimated, and the leakage volume is calculated from the total volume of water loss. The accuracy of the leakage estimation depends on the uncertainties associated the apparent loss estimation (Lambert et al. 2014; Lambert and Hirner 2000). UC estimation is essential for establishing the water balance and determining the water loss management performance indicators (AL-Washali et al. 2016; Alegre et al. 2006). As demonstrated in Chapter 7, UC remains the most uncertain component of the water balance. Estimating individual UC components is a tedious task that requires significant time and resources (AWWA 2009). There is no detailed reported case in the literature on UC estimation, and its complex and hidden nature leads to the frequent use of assumptions (AWWA 2016; Lambert and Taylor 2010; Mutikanga et al. 2011a; Seago et al. 2004). Regardless of whether the UC is a significant proportion of the water balance or not, UC estimation and associated uncertainty affect the leakage estimation, modelling, and management (Al-Washali et al. 2020b; Lambert et al. 2014). This impact is more critical when the proportion of UC is more significant, as in LMIC (Al-Washali et al. 2020b; Kingdom et al. 2006; Liemberger 2010; Liemberger and Wyatt 2018; Mutikanga et al. 2011a; Seago et al. 2004). The following section presents the methodologies of the five proposed approaches for UC estimation: (i) default values, (ii) water and wastewater balance, (iii) minimum night flow (MNF) analysis, (iv) component analysis of UC, and (v) correlating UC to disconnected customers.

9.2.1 Default values

Arbitrarily assuming the UC is the most common and straightforward method for UC estimation because of its associated difficulties and uncertainties. Table 9.1 presents common UC default values for high-, middle-, and low-income countries. A particularly low rate is assigned as the UC in high-income countries, ranging from 0.1% of the billed consumption in New Zealand to 0.25% of the system input volume in the USA (Austin Water Utility 2009; AWWA 2009; Jernigan 2014; Lambert and Taylor 2010; MDWSD 2011; Radivojević et al. 2008). For Europe, 0.2% of the billed consumption is proposed as the UC (Lambert et al. 2014). UC in LMIC is more significant than that in high-income countries. It ranges from 2% of non-revenue water (NRW) to 10% of the billed consumption. Seago et al. (2004) proposed assuming the UC of 2% to 10% of the NRW based on a subjective ranking of the UC level in the network. Water utilities classify the UC in their networks as very low, low, average, high, and very high based on their expectations. Although these assumptions appear to be close to the actual situation in LMIC, they are simply speculations and hence are not useful for monitoring the UC or improving leakage estimations. Whenever a water loss component is assumed to be a certain level, it cannot be monitored regularly.

Table 9.1. Overview of assumed values of UC

| High-income countries | | | Low- and middle-income countries | | | | | |
| | | | (Seago et al. 2004) | | (Mutikanga et al. 2011) | | (Liemberger 2010) | |
Country	UC	Study	Category	% NRW	Connections	% BC	Category	% BC
USA	0.25% of SIV	(AWWA 2009; Jernigan 2014; MDWSD 201)	Very low	2%			A1	<0.5%
Europe	0.2% of BC*	(Lambert et al. 2014)	Low	4%	< 5,000	0.5%	A2	0.5-1%
New Zealand	0.1% of SIV*	(Lambert and Taylor 2010)	Average	6%	5,000-50,000	2%	B	1%-2%
USA	0.3% of SIV	(AUW 2009)	High	8%	50,000-100,000	3%	C	2%-5%
Serbia	1% of SIV	(Radivojević et al. 2008)	Very High	10%	>100,000	10%	D	>5%

* Excluding exported water. SIV: system input volume; BC: billed consumption; and NRW: non-revenue water

Mutikanga et al. (2011a) proposed considering network size as a surrogate for the UC level in the network. They assumed a more significant UC for larger water networks and improved UC control for small networks. The UC is therefore assumed to exhibit a positive correlation with the number of service connections. Several factors affect the UC level, including socio-economic conditions, city size, culture, politics, and consumption patterns (Mutikanga et al. 2012). In fact, a water network in certain socio-economic contexts can have a different UC level from another similar-sized water network in a different context. The UC can be more significant in poorly managed small networks than that in larger networks with effective governance and management. Nevertheless, relating the UC level to the network size in LMIC remains reasonable. This is because larger networks in LMIC require higher UC management capacities, additional crews and detection teams, and more effective customer management, which is challenging in the LMIC context. Liemberger (2010) linked the UC level to the water utility management level. For this method, NRW management performance categories are employed. Category A1 includes outstanding utilities that can effectively manage NRW; thus, the UC is assumed to be at a very low level. In contrast, Category D utilities are the poorest in NRW management and hence are attributed very high levels of UC. The logic of this approach is intuitive; however, ranking water utilities and assigning certain UC levels remains a subjective process that does not inform sufficiently about the UC problem. While the proposed UC levels differ between various studies, assuming UC does not assist in its monitoring and control. "If you cannot measure it, you cannot manage it" (Peter Drucker 1909–2005). It is therefore impractical to assume that UC occurs at the same level when the UC itself must be addressed and reduced. When the UC level is significant and assumed in the network, reasonably accurate leakage estimations are not possible, making leakage estimation and minimisation a more ambiguous process.

9.2.2 MNF analysis

This method is typically used in a small part of the network or in a district metred area (DMA) to estimate network leakage based on analysing the night inflows into the DMA. Chapters 2 explains in detail this approach and Chapter 5 deals with its application. After estimating the leakage in the DMA, if the customer meters in the DMA were also read during the MNF testing, then the difference between the customer consumption in the DMA and the DMA inlet flow reading provides the NRW volume in the DMA. The apparent losses in the DMA can be estimated by deducting the leakage and any other unbilled authorised consumption in the DMA from the NRW volume. Knowing the metering and data acquisition error levels, the UC can be calculated from the total apparent loss using Equation 9.1.

$$AL = UC + CME + DAE \qquad (9.1)$$

where AL is the apparent loss; CME is the customer metring errors; and DAE is the data acquisition errors, which include the misestimation of unmetered consumption. Estimating the UC in several DMAs in the network provides an indication of the network UC level. Generalising the UC level estimated in the DMAs to the entire network depends on the representativeness of the DMAs to the entire network. If there are no representative DMAs, then this method is only applicable at the DMA scale.

9.2.3 Water and wastewater balance method

This method is elaborated in Chapter 5. It estimates the total volume of apparent loss without estimating the network leakage or UC components. Estimating the metering errors and data acquisition errors and deducting them from the total apparent loss give a direct estimate of the UC level in the network. The apparent loss can be estimated using this method via Equation 6.1 (Al-Washali et al. 2020b). Once the factors of Equation 6.1 are set, apparent loss monitoring can be conducted regularly. An example of Equation 6.1 after setting its factors is presented as Equation 9.2. Additional details on this method are presented in Chapters 5 and 7.

$$Q_{AL \; example \; system} = 1.07 Q_{ww} - 0.91 Q_{bc} \qquad (9.2)$$

where Q_{AL} is the apparent loss (m^3/yr); Q_{ww} is the inflow into the WWTP (m^3/yr); Q_{bc} is the billed consumption (m^3/yr). This method can be applied if the city has a centralised sewer system for some or all of the customers in the network. The results of this method can be generalised to the entire network if the sewer system covers a significant portion of the customers.

9.2.4 Component analysis of UC

The components of UC are presented in Table 9.2. The UC is estimated using this approach by quantifying the different types and components of the UC. There are two main UC components: UC by registered customers in the water utility databases and UC by unregistered users. Examples of UC by registered customers include (i) meter bypassing, (ii) meter tempering/meter fraud, (iii) adding an unregistered connection, (iv) illegally reconnecting to the water service after being disconnected, and (v) illegally reconnecting inactive customers. Examples of UC by unregistered users include (i) illegally established connections, (ii) unregistered consumption in informal settlement and low-income areas, (iii) unauthorised use by small informal businesses such as car washers or open-air restaurants, (iv) unauthorised use by construction companies, and (v) water theft from network equipment such as fire hydrants. These types of UC should be estimated for domestic, commercial, industrial, governmental, and agricultural uses. Agricultural UC rarely occurs, but when it does, it consumes massive amounts of water and causes considerable apparent losses for the water utility.

Table 9.2. UC Components

Unauthorised Consumption (UC)			
Registered Customers		Unregistered Consumers	
Reported	Unreported	Reported	Unreported
Domestic	Domestic	Domestic	Domestic
Commercial	Commercial	Commercial	Commercial
Industrial	Industrial	Industrial	Industrial
Governmental	Governmental	Agricultural	Agricultural
Types: meter bypass, meter fraud, additional unregistered connection, disconnected but illegally reconnected, inactive customers illegally reconnected.		Types: illegal connections, unregistered consumption in low-income areas, informal traders, water theft from hydrants or other network equipment.	

With this approach, the UC value should be quantified for each UC component using Equation 9.3. UC by the total registered customers, reported registered customers, and unreported registered customers can be estimated using Equations 9.4, 9.5, and 9.6, respectively. Similarly, the UC by unregistered consumers, reported unregistered consumers, and unreported unregistered consumers can be estimated using Equations 9.7, 9.8, and 9.9, respectively. The total UC volume can then be calculated by summing the UC by the registered and unregistered consumers as shown in Equation 9.10.

$$UC = \sum q_i \times n_i \qquad (9.3)$$

$$UC_{reg} = UC_{reg,rep} + UC_{reg,unrep} \qquad (9.4)$$

$$UC_{reg,rep} = q_{d,rr} \times n_{d,rr} + q_{c,rr} \times n_{c,rr} + q_{ind,rr} \times n_{ind,rr} + q_{gov,rr} \times n_{gov,rr} \qquad (9.5)$$

$$UC_{reg,unrep} = q_{d,ru} \times n_{d,ru} + q_{c,ru} \times n_{c,ru} + q_{ind,ru} \times n_{ind,ru} + q_{gov,ru} \times n_{gov,ru} \qquad (9.6)$$

$$UC_{unreg} = UC_{unreg,rep} + UC_{unreg,unrep} \qquad (9.7)$$

$$UC_{unreg,rep} = q_{d,ur} \times n_{d,ur} + q_{c,ur} \times n_{c,ur} + q_{ind,ur} \times n_{ind,ur} + q_{agr,ur} \times n_{agr,ur} \qquad (9.8)$$

$$UC_{unreg,unrep} = q_{d,u} \times n_{d,u} + q_{c,u} \times n_{c,u} + q_{ind,u} \times n_{ind,u} + q_{agr,u} \times n_{agr,u} \qquad (9.9)$$

$$UC_{total} = UC_{reg} + UC_{unreg} \qquad (9.10)$$

where q_d is the average domestic customer consumption (e.g. m^3/yr); q_c is the average commercial customer consumption; q_{ind} is the average industrial customer consumption;

q_{gov} is the average governmental customer consumption; q_{agr} is the estimated average irrigation quantity for illegal agricultural users in the city; and n_d, n_c, n_{ind}, n_{gov}, and n_{agr} are the estimated numbers of reported and unreported unauthorised users for domestic, commercial, industrial, governmental, and agricultural uses, respectively.

Quantifying UC using this approach is clearly challenging as there are many different UC types and components. Estimating the average domestic and non-domestic water consumption from the water utility billing data is straightforward, but this step can have uncertainties associated with generalising the average consumption. However, the critical step in this approach is to determine the number of illegal users in each component. Unreported illegal use is expected to be a major proportion of UC, and estimating this particular figure is the main limitation of this approach. Water utilities should improve their UC estimates over time. However, the results of this approach are primarily based on the UC that the utility detects or perceives, but they do not include all of the UC (unknown UC). This implies that this method is susceptible to considerably underestimating the UC.

9.2.5 Correlation of UC to disconnected customers

Another possible method for estimating the UC is by using the level of disconnected customers as a surrogate of the UC level in the network because disconnected customers who do not apply for re-connection for long time periods tend to turn into illegal users. Firstly, the UC should be estimated using one of the aforementioned methods. Secondly, the UC volume should be converted to an equivalent number of illegal cases in the network using Equation 9.11. The UC factor (F_{UC}), which relates the level of UC in the network to the number of disconnected customers, can then be established using Equation 9.12. Knowing the F_{UC}, the UC can be directly estimated based on the number of disconnected customers using Equation 9.13.

$$n_{eqv} = \frac{UC}{q_d} \qquad\qquad (9.11)$$

$$F_{UC} = \frac{n_{eqv}}{n_{disc}} \qquad\qquad (9.12)$$

$$UC = F_{UC} \times n_{disc} \times q_d \qquad\qquad (9.13)$$

where n_{eqv} is the equivalent number of illegal cases in the network, UC is the unauthorised consumption (m³/yr), q_d is the average consumption per customer (m³/yr), F_{UC} is the UC factor, and n_{disc} is the number of customers who remain disconnected from the water service. In this method, it is assumed that the UC is positively correlated with the number of disconnected customers in the network. This is a valid assumption because the utility customer management approach affects the UC level in the network to a large extent. When laxity and poor customer management is the practice by the water utility, nonpayers and disconnected customers in the network increase, which in turn increases the UC level. Although this method is simple and straightforward, its drawback is the prerequisite of estimating the UC in the network using a different approach or assuming the F_{UC} based on data of another case study, the accuracy of which remains critical.

9.3 APPLICATION OF UC ESTIMATION METHODS

The five methods were applied to six different cases in Africa and Asia: (i) Zarqa, Jordan, a water network with 160,000 customers, an intermittent supply with an average supply time of 36 h/w, a 33-m average pressure, and an NRW volume (2014) of 43.7 million cubic meters (MCM)/yr, which was 65% of the system input volume (SIV); (ii) Sana'a, Yemen, a water network with 94,723 customers, an intermittent supply with an average supply time of once per week for 4.4 hr/d, a 10-m average pressure, and an NRW volume (2009) of 8.6 MCM/yr, which was 39% of the SIV; (iii) Mwanza, Tanzania, a water network with 49,284 customers, an intermittent supply with a daily supply for 22 hr/d, a 58-m average pressure, and an NRW volume (2015) of 13.9 MCM/yr, which was 46% of the SIV; (iv) Utility A, Africa, a water network with 511,718 customers, a continuous water supply, a 55-m average pressure, and an NRW volume (2016) of 91.7 MCM/yr, which was 26% of the SIV; (v) Utility B, Africa, a water network with 196,670 customers, a continuous water supply, a 50-m average pressure, and an NRW volume (2016) of 20.3 MCM/yr, which was 50% of the SIV; and (vi) Utility C, Africa, a water network with 652,217 customers, a continuous water supply, a 62-m average pressure, and an NRW volume (2016) of 221.8 MCM/yr, which was 38% of the SIV. The African networks were labelled as such to protect the data of these utilities.

9.3.1 Default values (M₁)

The default values of UC based on the NRW volume, billed consumption, or SIV can be straightforwardly applied to any water system to estimate UC. Figure 9.2 shows the UC for each water network based on the proposed values for LMIC (Liemberger 2010; Mutikanga et al. 2011a; Seago et al. 2004). There are significant differences between the assumptions. Figure 9.2 demonstrates that the differences between the default assumptions become more significant for larger networks. Furthermore, with these assumptions the UC level is always the same over time and do not provide any information regarding the actual UC in the field or the progression of its management.

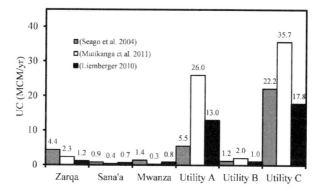

Figure 9.1. UC volume based on common default assumptions for the six water networks

9.3.2 MNF analysis (M$_2$)

This method was applied in the Zarqa and Mwanza water networks as shown in Table 9.3. The DMA was not representative of the entire network for either case. For Zarqa, the UC volume estimated using this method was larger than the total billed consumption (Table 9.3). This indicates the UC can be considerably overestimated. The water loss level in Zarqa was abnormally high, and there was illegal water use in the network for irrigation purposes in the suburbs of the city. One illegal user irrigating a farm in the suburb causes more revenue losses to the water utility than hundreds of domestic illegal users in the city centre. However, the UC level in Zarqa should not be more than the water sold by the water utility. Therefore, the average UC value using this method and the default value (10% of the billed consumption) were calculated for this network. The later provided a reasonable UC level estimate that was closer to what was expected by the Zarqa water utility specialists. The MNF analysis was also applied to Mwanza, Tanzania, and the DMA was also not representative; the UC volume appeared to be substantially underestimated. Therefore, UC level estimation by MNF in a DMA is unreliable. These results support combining UC estimations via an MNF analysis with other methods (Mutikanga et al. 2011a).

9.3.3 Water and wastewater balance (M$_3$)

The water and wastewater balance method was applied in Sana'a and Mwanza, as demonstrated in Chapters 6 and 7. The customer meter errors and data acquisition errors were estimated to be 0.6 MCM/yr and 0.8 MCM/yr for Sana'a, and 0.8 MCM/yr and 0.8 MCM/yr for Mwanza, respectively. The UC volume was then calculated based on the total apparent loss volume. Table 9.3 displays the water and wastewater balance methodology results. There were significant differences between the default value and

179

the water and wastewater method estimations. There is no proven methodology to validate these results. However, based on discussions with the utility specialists, these results were reasonable and representative.

Table 9.3. UC estimation (m³/yr) using different methods

Method	Utility	SIV	BC	NRW	NRW	UAC	RL	CMEs	DHEs	UC
MNF Analysis	Zarqa	67	23	44	65.3%	1.4	16	2.6	-0.39*	24.1
	Zarqa**	67	23	44	65.3%	1.4	27	2.6	-0.4	13.2
	Mwanza	30	16	14	46.4%	0.1	12	0.8	0.8	0.1
W & WW Balance	Sana'a	22	14	9	38.8%	0.1	3	0.6	0.8	4.4
	Mwanza	30	16	14	46.4%	0.1	7	0.8	0.8	5.7
Component Analysis of UC	Utility A	352	260	92	26.1%	3.7	70		13.3	4.3
	Utility B	41	20	20	49.9%	0.3	16		2.6	1.2
	Utility C	578	357	222	38.4%	73.2	107		28.9	12.7

* DHEs are overbilled consumption ** UC based on average of MNF and default value of 10% of BC.

UAC: unbilled authorised consumption; RL: real loss; CMEs: meter errors; DHEs: data handling errors

9.3.4 Component analysis of UC (M₄)

The component analysis of UC was applied to the three utilities in Africa. The reported and unreported UC were estimated for the registered and unregistered customers of the utility. For the registered illegal users, the reported (detected) number of illegal use cases was multiplied by the average consumption, based on the utility records, for both domestic and non-domestic users. The unreported UC for the registered customers was estimated based on the estimated number of customers who were disconnected from the water service, never applied for re-connection, and remained disconnected. The UC component estimates for the registered customers are shown in Table 9.4. Similarly, the reported UC by unregistered users was estimated based on the records of the utility for detected illegal users who have no records in the utility. The unreported UC by unregistered users was estimated based on a combination of methods. The UC by informal (low-income) settlements was estimated based on the records of informal settlements without water service and the monthly average consumption for a single settlement. UC from fire hydrants to wash cars in the street was estimated by collecting secondary data of these small businesses and conducting field visits to four car washes to determine the average water consumption of the car wash per day. UC used for construction work was estimated based on available records in the utility for 1,641 single-story houses that were built without water approval and therefore used water illegally from the fire hydrants. The average consumption of single-story houses based on the utility records was used to provide a UC estimate for construction work. UC used by open-air restaurants were estimated based on secondary data on this small business and three restaurant field investigations into the average consumption per restaurant. The field and collected data were then presented to and discussed with the utility specialists in three focus group discussions with 13 people in total. Table 9.4 presents the estimated UC components for Utility A. The detailed UC estimations for Utility B and Utility C are presented in the supplementary data of this chapter, and the total estimated UC is presented in Table 9.3. The estimated UC for Utility A using this method was 4.3 MCM/yr, which was less than the estimated UC by the utility based on the default assumption of 6% of NRW, which was 5.5 MCM/yr. For Utility B, this method yields the same results as those of the default assumption used by the utility. For Utility C, this method estimated the UC in the network at 12.7 MCM/yr while the default assumption for this utility was 10% of the NRW volume, which was equivalent to 22.2 MCM/yr. This method significantly underestimated the UC level for Utility A and Utility C. This is because the UC component analysis only estimates the known parts of the UC due to the hidden nature of UC, which cannot be fully detected or recognised by the water utility.

Table 9.4. UC component analysis in Utility A, Africa

		Registered illegal users			
UC	Type	Use	Cases	q_{avg} (m³/yr)	UC (m³/yr)
Reported 135	Detected	Domestic	126	391	49,090
		Non-domestic	9	1,998	18,881
Unreported 6720	Disconnected*	Domestic	6,250	391	2,443,594
		Non-domestic	470	1,998	939,859
		Total	6,855		3,451,424
		Unregistered illegal users			
UC	Type	Use	Cases	q_{avg} (m³/yr)	UC (m³/yr)
Reported 8	Detected	Domestic	7	391	2,909
		Non-domestic	1	1,998	1,119
Unreported 9,846	Suspected	Informal settlements	8,121	72	584,712
		Car wash	84	329	27,636
		Construction works	1,641	160	262,560
		Total	9,854		878,936
				Grand Total	4,330,360

9.3.5 Correlation of UC to disconnected customers (M₅)

This method was proposed to provide a quick snapshot estimate of the UC level as an alternative to using the default UC value based on utility information as a surrogate for the UC level in the network. The UC estimations by the above four methods were used to generate the F_{UC} via Equation 9.12, as shown in Table 9.5. The F_{UC} can be used in Equation 9.13 to provide a quicker and more objective UC estimate than those of the default values while also indicating the changes in the UC level in the system over time.

Table 9.5. Estimating F_{UC} for the six case studies

Case study	UC (Mm³/yr)					q_d (m³/yr)	$n_{equ.}$	$n_{dis.}$	F_{uc}
	M_1	M_2	M_3	M_4	Used				
Zarqa	4.4	24.1			14.2	145	98,227	20,094	4.9
Sana'a	0.9		4.4		4.4	144	30,223	15,740	1.9
Mwanza	1.4	0.1	5.7		5.7	327	17,344	4,304	4.0
Utility A	5.5			4.3	5.5	391	14,070	6,720	2.1
Utility B	1.2			1.2	1.2	141	8,647	8,007	1.1
Utility C	22.2			12.7	22.2	665	33,359	7,191	4.6

The F_{UC} values in Table 9.5 range between 1.1 and 4.9. To generalise the F_{UC} range, more cases must be analysed. However, based on the analysis of six utilities in developing countries, Table 9.6 displays the proposed values of F_{UC} for the initial UC estimation. Estimating UC using the matrix in Table 9.6 is possible based on two factors, the NRW level as proposed by Liemberger (2010) and network size as proposed by Mutikanga et al. (2011a). The range of each specific item in the matrix allows for insights into the water utility on the UC level as used by Seago et al. (2004). The matrix values in Table 9.6 were based on the case study data. Table 9.7 presents the NRW level and network size ranges, to identify case-specific NRW level and network size. Based on Table 9.5, the average F_{UC} of the four large-sized utilities (Zarqa and Utilities A, B, and C) with high NRW level was 3.2. As the UC in three out of the four utilities was underestimated, the average of these four utilities was also expected to be underestimated. The F_{UC} range for this category was therefore created with a larger bound of 4. This upper bound for large-sized utilities was then redistributed in the same row in Table 9.6 for the average and small NRW level ranges. Similarly, the average F_{UC} for medium-sized utilities with high NRW levels was 3, and this figure was redistributed for the small and average NRW level categories. There were no small utilities in the analysed cases, but a lower bound was provided as UC tends to be less significant when the network size is smaller (Mutikanga et al. 2011a).

In summary, when estimating the UC using this method, the F_{UC} should first be estimated for the specific utility using Equation 9.12. If no data is available, then the proposed F_{UC} can be used for the initial UC estimation. The network size and NRW level in the water utility should be defined using Table 9.7, and then the proposed F_{UC} value can be obtained from Table 9.6. Afterwards, the UC volume can be estimated using Equation 9.13.

Table 9.6. Proposed matrix for estimating UC using F_{UC}

		NRW Level		
		Low	Average	High
Size of Network	Small	0.5-1	1-2	>2
	Medium	1-1.5	1.5-3	>3
	Large	1-2	2-4	>4

Table 9.7. Network size and NRW level ranges

Size of Network		Small	Medium	Large
	Connections	<20,000	20,000-150,000	>150,000
Level of NRW		Low	Average	High
	ILI	1-4	5-8	>8
	%	5-10%	10-25%	>25%

The impact of UC estimations on water loss management was investigated for three of the six analysed networks, Zarqa, Sana'a, and Mwanza, using the demonstrated methodology in Chapter 7. The estimation of leakage and benefits of leakage reduction interventions were made using a leakage economic modelling tool (Sturm et al. 2014). For the Zarqa network, the leakage volume was estimated to be 37.85 MCM/yr based on using default values for the UC. The MNF analysis resulted in a leakage volume of 16.0 MCM/yr, and the average of these two methods was 26.9 MCM/yr, as shown in Table 9.3. The difference between the two estimates is significant, and consequently, the annual water loss cost varied from 12.0 million USD to 21.5 million USD. Modelling the leakage intervention benefits, the annual potential benefits of a pressure reduction in Zarqa by one bar (from 3.3 bars to 2.2 bars) was 1.2 million USD for the MNF method, 2.7 million USD for the default assumption method, and 1.9 million USD for the average of both methods. Similarly, for Sana'a, the UC was estimated through the default assumption and the water and wastewater balance method, and the leakage was estimated to be 5.8 MCM/yr and 2.8 MCM/yr, respectively. The annual water loss cost varied from 12.0 million USD to 21.5 million USD. Analysing the feasibility of reducing the pressure in the network by 0.2 bar (from 1 bar to 0.8 bar), the estimated benefits based on the default assumption and the water and wastewater method were 0.4 million USD and 0.2 million USD, respectively. Similar results were also found for Mwanza for the three methods: the default assumption, MNF analysis, and water and wastewater method. The water loss cost ranged from 3.3 million USD to 3.6 million USD, and the potential benefits of a pressure reduction by 2 bars (from 5.8 bars to 3.8 bars) varied from 0.5 million USD to 1.0 million USD. These results confirm the UC estimation sensitivity and indicate that estimating the UC accurately is essential for water loss management, especially in LMIC where UC is common.

9.4 CONCLUSIONS

This study analysed five possible methods for estimating UC, which is a critical component of the water balance and significantly affects the analysis of other water loss components including leakage rate and reduction planning. Assigning default values is the most common approach for UC estimation due to the complexity of the UC components and their hidden nature. These assumptions are both arbitrary and critical for water loss analysis. The assumptions always display the same UC level over time, providing no information on the actual UC levels in the field and no indication of the progression of UC management. The results showed that generalising the UC level for a network based on field measurements using MNF analysis in one (or several) DMAs is unreliable. This approach is also not reasonable because the UC level varies within the network based on the socio-economic conditions. Unlike the default values, the water and wastewater balance method objectively estimated the UC volumes in two case study systems, and this method enables the regular tracking of UC levels in the network. The accuracy of UC estimated using this method depends on the accuracy of the apparent loss volume estimations in the network and the existence of central sewers for part or all of the water network. Quantifying the individual UC components is theoretically possible, and equations to estimate the UC components were presented. However, the critical step in this approach is to determine the number of illegal users in each component, especially illegal users that are not registered in the water utility. The UC from farm irrigation in the suburbs and semi-urban areas is extremely sensitive. One illegal user in the suburbs irrigating a farm causes greater revenue losses than hundreds of domestic illegal users in the city centre. Tracking the UC of all registered and unregistered users who use water from the network for different domestic, commercial, and agricultural uses is challenging. The UC estimated using this approach is susceptible to underestimation because water utilities cannot recognise or detect all illegal uses in the network. UC estimation is clearly a dilemma that water utilities should address. Water utilities can estimate UC using the water and wastewater balance method and then break down the UC into components using the component analysis of UC. This approach enables water utilities to estimate the UC volume and understand its types and components, which is more effective for UC management. Alternatively, if this process cannot be applied, water utilities can use UC correlation to the disconnected customers, as proposed in this chapter, to analyse and track the UC level in the network.

9.5 SUPPLEMENTARY DATA

Table 9.8. Component analysis of UC for Utility B, Africa

Registered illegal users

UC	Type	Use	Cases	q_{avg} (m³/yr)	UC (m³/yr)
Reported	Detected 0	Domestic	0	0	0
		Non-domestic	0	0	0
Unreported	Disconnected* 8,007	Domestic & Non-domestic	8,007	141	1,130,024
		Total 8,007			1,130,024

Unregistered illegal users

UC	Type	Use	Cases	q_{avg} (m³/yr)	UC (m³/yr)
Reported	Detected 25	Domestic	0	0	0
		Non-domestic	25	98	2,450
Unreported	Suspected 646	Informal settlements	360	72	25,920
		Car wash	143	190	27,170
		Open restaurants	143	7	1,001
		Total 671			56,541
		Grand Total			1,186,565

Table 9.9. Component analysis of UC for Utility C, Africa

Registered illegal users

UC	Type	Use	Cases	q_{avg} (m³/yr)	UC (m³/yr)
Reported	Detected 7,630	Domestic	7,020	665	4,668,034
		Non-domestic	610	2,578	1,573,611
Unreported	Disconnected* 7,200	Domestic	6,624	665	4,404,960
		Non-domestic	576	2,578	1,484,928
		Total 14,830			12,131,533

Unregistered illegal users

UC	Type	Use	Cases	q_{avg} (m³/yr)	UC (m³/yr)
Reported	Detected 487	Domestic	448	665	297,947
		Non-domestic	39	2,578	100,439
Unreported	Suspected 2,711	Informal	2,645	72	190,440
		Car wash	33	548	18,068
		Open restaurants	33	13	434
		Total 3,198			607,327
		Grand Total			12,738,860

185

10

CONCLUSIONS AND FUTURE OUTLOOK

10.1   Water loss assessment in intermittent supply

Water losses in distribution networks are high at the global level and higher still in developing countries. This is despite great efforts by water utilities, technological breakthroughs, and intensive national and donor-assisted projects to manage water losses in distribution networks. This situation requires reflecting on the concrete challenges that water utilities face in developing countries as well as re-thinking the available methods and tools to manage and control the losses in their networks. The first, and probably the most important, is the diagnosis and assessment of water loss and its components in the network. Water utilities in developing countries face different problems compared to what is common in developed countries. In addition to the high water loss level, these utilities encounter unique complications, the most important of which are the common intermittent supply, unauthorised use (also for irrigation) and "kleptomania" from water networks, poorly designed and constructed networks with a majority in dilapidated condition, problems of data accuracy and availability, and often an increasing scarcity of water resources. In light of the above, the assessment of the water loss level and components in intermittently operated networks in developing countries remains a difficult problem. This study attempted to improve water loss assessment in intermittent supplies by analysing this critical subject in detail and developing several tools and methods so that a more effective water loss management strategy may be set, monitored, and fulfilled.

The first chapter of this thesis presented the research problem and the objectives of the study, Chapter 2 reviewed the state-of-the-art water loss assessment methods, and Chapter 3 dealt with the water loss assessment software tools. Chapter 4 addressed the assessment of the total loss in intermittently operated networks, Chapter 5 examined specifically the applicability of MNF analysis in a DMA in an intermittent supply, Chapter 6 presented the development and application of a new water loss component assessment method, and Chapter 7 compared the applications, uncertainties and implications of the different water loss assessment methods, including the developed method in Chapter 6. Chapter 8 analysed the customer water meter performance in an intermittent supply, and finally, Chapter 9 dealt with the unauthorised consumption in the network. This chapter presents the conclusions related to what has been discussed in the different chapters. It begins with the conclusions related to assessing the total water loss in the network (Chapter 4) and then deals in detail with water loss component assessment methods with a focus on apparent losses. Finally, this chapter closes this thesis with a future outlook of water loss assessment in networks with intermittent supply.

10.2 MONITORING WATER LOSS PERFORMANCE IN INTERMITTENT SUPPLY

The results revealed that all the NRW performance indicators, without exception, are not meaningful when the water supplied into the network is variable, as is the case in intermittent supply networks, unless these performance indicators are normalised. The volume of NRW varies monthly and annually according to fluctuations in water production. An increase in the NRW level does not necessarily indicate worse performance, as it could be due to an increase in the amount of supplied water, and similarly, a decrease in the NRW level does not necessarily indicate a better performance, as it could be due to a decrease in the supplied water; *the NRW level should therefore be normalised.* Normalisation is not a completely new concept, it is commonly recommended to normalise the performance indicators of real losses, through the "when system is pressured" (w.s.p.) adjustment. Extending this approach for normalising the NRW and apparent losses is, therefore, a practical approach. However, several limitations of this approach were reported in Chapter 4. Normalising the volume of apparent losses through this method is susceptible to significant overestimation of the apparent losses. However, the most critical limitation of this approach is the questionable (normalised) results when the average supply time in the network is low or close to zero (Figure 10.1). In the water supply system of Sana'a, the normalised NRW volume began to change dramatically when the supply time was less than eight hours per day. Similar graphs are expected for other networks because the shape of the beginning and end of the curve in Figure 10.1 will be the same for each intermittent water supply system. Further investigation should be carried out to confirm this, however in all cases a point should exist where the curve can be divided into linear and non-linear parts. In order to use this approach, the normalisation factor curve (e.g., Figure 10.1) should be linearised. Standard linearisation factors can be published in updated popular benchmarking references (IWA and AWWA), or developed at national or utility level.

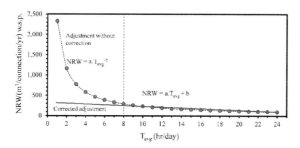

Figure 10.1. Example of a linearised NRW normalisation curve using the w.s.p. adjustment

The linearised w.s.p. adjustment approach can then be used for internal target setting and monitoring, as well as for benchmarking and comparing different water utilities with different levels of supplied water. Alternatively, the volume of NRW can be normalised for intermittent supply using linear regression analysis, as discussed in Chapter 4. This approach is practical and more reliable for reflecting the actual progression and regression of NRW management. Regression analysis can be used at the utility level for target setting, NRW monitoring, and the evaluation of NRW management interventions. Comparing the performance of different utilities is not possible with this approach, and the linearised w.s.p. adjustment remains the only method for benchmarking intermittently operated networks.

10.3 LIMITATIONS OF THE AVAILABLE METHODS AND TOOLS FOR NRW COMPONENT ASSESSMENT

10.3.1 Top-down water balance[9]

The top-down water balance is a promising tool for assessing losses in water networks and is the most common method of estimating the leakage volume for the whole network. What distinguishes this method is that it is cost-effective, does not depend on extensive fieldwork, and does not require estimating the average pressure for the entire network - which is difficult to determine. In contrast, this method necessitates, first and foremost, the quantification of apparent losses in the network, and thus the accuracy of the estimated volume of real losses depends on the accuracy of the estimation of the apparent losses. However, the assessment of apparent losses in intermittent supplies in developing countries (which are usually very significant) is a subject that has not received sufficient research attention and is therefore beset with uncertainties, which has a major influence on the accuracy of this method. This is due to the fact that the adopted method for calculating the weighted error of customer meters has not yet taken into account the actual flows that pass through the water meter but the typical customer consumption flow profile. This approach affects the accuracy of the estimated errors of customer meters and in turn influences the accuracy of the top-down water balance. But more importantly, estimating the level of unauthorised consumption in the network remains the principal challenge. Based on the analysis of the data from six case studies in this research (in addition to seven other case studies that were analysed and not included in this study, as well as many years of professional experience), the largest components of water losses are leakage and unauthorised consumption. In fact, these are the most significant and important

---

[9] The top-down water balance has been reviewed in Chapter 2 and applied in three cases and compared with other methods in Chapter 7.

components of water loss in the analysed cases (and probably many other cases). Customer meters' inaccuracy was comparatively less significant in the analysed cases as it becomes critical only in combination with water tanks, float valves and continuous supply. In this case, the accuracy of the meters becomes critical, whereas it is most likely not as important in other situations. As unauthorised consumption is the second most important component of water loss, estimating the volume of unauthorised consumption remains the cornerstone of obtaining accurate results through the top-down water balance method. Adopting arbitrary assumptions is extremely convenient, but is neither scientific nor practical. This is evident from the results of Chapter 7, as this method was shown to consistently and significantly underestimate the volume of apparent losses and overestimate the leakage volume.

10.3.2   Minimum night flow analysis[10]

Estimating the leakage volume in a DMA based on the field inflow measurements is a method that is mainly useful at the DMA scale. Chapter 5 highlighted that analysing MNF data collected over one day and deducing the leakage volume in the DMA does not appear to be a valid option in the case of intermittent supply. Rather, the DMA must be temporarily transformed into a continuous system by supplying water continuously for several days until all the ground and elevated tanks in the network are guaranteed to be filled with water. If this is the situation, the results should be reliable and the MNF curve of the inflow into the DMA (e.g., Figure 5.4) will begin to repeat the same or closer measurements. This procedure is still applicable in principle in intermittent supply, but for specific or occasional tests. The application of this method on a regular or permanent basis is, in fact, not possible in the case of intermittent supply because it requires disturbing and adjusting the schedule of water distribution in the network. Hence, permanent or regular monitoring of the leakage volume in DMAs still requires other methods and tools. An additional problem also arises with this method. The MNF is not necessarily exclusively a night flow, but may also represent a day flow, as in the case of the Zarqa water network where the minimum flows occurred between 12:15 AM and 7:00 AM. This imposes a further challenge, as the customer consumption during the MNF may not always represent the lowest consumption expected while customers are asleep, but may represent a significant consumption as some customers are already awake. Estimating the customer consumption at the time of the minimum flow therefore remains a potential problem facing the application of this method in intermittent supply. This could lessen the accuracy of the method, or undermine the basis of the method. Ultimately,

---

[10] The minimum night flow analysis has been reviewed in Chapter 2, applied in Chapter 5, and further applied and compared with other methods in Chapter 7.

the customer consumption during the minimum flow depends not only on each individual DMA but also on each individual test.

Once the customers' consumption and the leakage volume at the time of the minimum flow have been determined, the approach of modelling the leakage throughout the day remains subject to the outcomes of a current active research. The leakage–pressure relationship is a problem of mathematical and empirical scientific investigation. Certainly, the Torricelli equation failed to model the relationship between pressure and leakage in networks containing plastic and non-rigid pipes. Modified equations have therefore been proposed, taking into account the variable area of the leakage, which is affected by pressure to a large extent, but the empirical FAVAD principle remains the most commonly applied. FAVAD is a very practical model, but it contains simplifications that affect the accuracy of the calculated leakage volume. The most important limitation in this principle is that the leakage exponent (N_1) itself is affected by varying pressures throughout the day, and there is therefore a need to convert the proposed mathematical models into practical solution for practitioners in the field. This matter also affects the accuracy of the 'minimum flow' method.

10.3.3   BABE analysis[11]

The third method for leakage assessment is to analyse the leakage components in the network (system-wide) using BABE analysis. The most important aspect of this method is that the leakage volume depends on two main factors, the leak flow rate and the leak run time. This method clearly reflects that the leak run time is more important and more sensitive than the leak flow rate. In this sense, leakage in the form of large or giant bursts (which are often repaired within a short period of time) constitutes a much smaller volume of water loss than that associated with small hidden leaks that do not manifest on the surface but run for a long time. This method helps water utilities understand the leakage components and recognise the impacts of utility policies on the leakage level. However, this method analyses a fraction of the leakage in the network, as in the case of the Zarqa network, where BABE analysed only 26% of the leakage volume. In two case studies in this research, an automatic, effective and reliable documentation system for network bursts existed, therefore all the bursts above the ground surface are believed to have been documented. However, this method analysed only a small fraction of the total leakage volume despite the use of a high (2.5) infrastructure condition factor (ICF), and despite keeping the flow rates in the model the same as in developed countries (over-estimated). This method underestimates the leakage volume due to the inherent assumptions in the

---

[11] The BABE analysis has been reviewed in Chapter 2, applied in Chapter 5, and further applied and compared with other methods in Chapter 7.

BABE equations related to the unavoidable background leakage, which are deduced from case studies with completely different characteristics, construction quality, and utility policies (related to leakage detection). An additional shortcoming of this method is the use of the average pressure of the entire network, which is difficult to determine as an annual (spatial and temporal) average with sufficient accuracy. Although this method is applicable in intermittent supply networks, it should not be relied upon for estimating the total leakage volume, even if a high ICF is used. The method is useful and practical for analysing and understanding the nature of the leakage components and indicating the great importance of hidden leaks relative to the bursts manifesting on the surface.

In confirmation of the above, Chapter 7 revealed that the process of water loss component assessment is marred by many uncertainties, which in turn affect the effective planning and economic feasibility of water loss reduction, in particular, evaluating the economic feasibility of leakage detection surveys, but also pressure management and other loss reduction options. There is no way around this, except by verification of the water loss component assessment. In this regard, uncertainty analysis is an effective tool for improving the output of the method because it clearly indicates which inputs should be double-checked as a priority in order to obtain more reliable results.

The limitations listed above regarding the methods of assessment of the water loss level and components were naturally reflected in the available software tools for water loss management. There is great interest in, and a corresponding growing number of software tools for, assessing and reducing water losses, many of which are free and available to all, including water utilities with intermittent supply. Chapter 3 reviewed these tools in detail and discussed the aspects that should be covered in future versions of these tools. However, two main points should be highlighted:

(i) Firstly, taking intermittency into account in the inputs of water loss assessment software tools expands, to a great extent, the beneficiaries of these tools. For example, there are 3,255 water utilities in India alone, many of which represent an intermittent supply, and some of these utilities serve a population three to four times that of the Netherlands. Consideration of intermittency in water loss assessment tools is therefore important and potentially very beneficial. If this is the case, then the normalisation of water losses within these tools should also be taken into account. In this context, normalisation does not only indicate the normalisation of real losses but, more importantly, the normalisation of the total volume of water losses in the network.

(ii) Secondly, while various tools exist for assessing, modelling, and planning the reduction of real losses, (free) software tools for assessing and planning the reduction of apparent losses are almost non-existent. In an intermittent supply context, the assessment and management of apparent losses have not yet been sufficiently researched, let alone adequately recognised in water loss management tools. Further attention and research in the field of apparent losses is highly necessary, specifically in intermittently operated networks.

10.4   FOCUSSING ON APPARENT LOSSES

Chapter 6 describes a proposed method for estimating the apparent losses in the network using the apparent loss estimation (ALE) equation, based on the establishment of the water and wastewater balance. This method utilises several inputs with limited sensitivity, but mainly relies on two types of measurements: (i) billed consumption measurements and (ii) measurements of the WWTP inflows. If a meter with a good accuracy (e.g., an ultrasonic meter with four-path crosswise measurements; ±0.5–1%) is installed at the inlet of the WWTP, the apparent losses can be assessed and monitored regularly by calculating the ratio between the per-customer WWTP inflow to the per-customer billed consumption. By monitoring this ratio, the apparent losses in the network can be deduced straightforwardly on a regular basis. This is performed after the estimation and optimisation of the factors of the ALE equation (Equation 6.1) for each specific network, which, by the way, are factors with low sensitivity to the output of this method. If in the future all customers have smart meters that send daily or real-time consumption data, there will exist the possibility for real-time monitoring of apparent losses, and thus real losses, in the network. The above demonstrates the potential of the proposed method. Nevertheless, this method has not been developed as a one-size-fits-all method, but rather has been developed for cities that have a high level of apparent losses, seasonal rainy days, and central sewerage networks (to which the majority of the water utility customers are connected). In such cases, estimating the apparent losses, and thus the real losses, regularly, systematically, and cost-effectively is possible using this method, without the need for night field measurements with a disturbed distribution schedule or assumptions with high sensitivity as in the other methods. After quantifying the total volume of apparent losses in the network, the next step is to break down the apparent losses into subcomponents. If the losses resulting from data acquisition and billing errors are assessed and the losses resulting from the customer meter inaccuracy are estimated, it then becomes fairly straightforward to calculate and objectively monitor the volume of unauthorised consumption in the network.

Chapter 8 demonstrates the methodology for calculating the weighted error of customer meters in the case of intermittent supply. This is not achieved by relying on customer consumption flow rates, but by recognising the flow rates of the float valves (FV) in the tanks. These FV inflow rates are lower than the consumption flow rates (tank outflows) owing to the balancing effect of the tank itself. In addition, it should be noted that the customer meter inaccuracy is not so critical in intermittent supplies. When the network supply time is low (e.g., less than eight hours per day; i.e., < 8/7), the weighted error of the customer meters may become positive (over-registration) because the associated intrinsic error of the meter is originally positive at certain flows, which are frequently observed in this situation. This also does not consider the air that enters the intermittently operated network and then exits the network through the water tanks when the supply is resumed. This air passes through the mechanical meters which record it as a quantity of water, thus increasing the probability of over-registration of the network customer meters. Conversely, the customer meter inaccuracy becomes a critical matter when intermittent supply networks are transformed into continuous supply networks with water tanks (and FVs) remaining in the network. In this case, the tanks often become full and the impact of customer consumption on the water level in the tank is small, causing the float valve to open slightly, and introducing low flows that significantly deteriorate the accuracy of the meter. This matter should be taken into account when shifting from intermittent to continuous supply, by adjusting the meters replacement policy to accommodate a lower starting flow rate of customer meters, as required for each specific network.

After assessing the customer meter inaccuracy as well as losses related to data acquisition and billing errors, the remainder of the apparent losses corresponds to the amount of unauthorised consumption in the network. With this methodology, it is possible to estimate the unauthorised consumption in the network in a more reliable way than by utilising "blind" assumptions (default values) that do not rely on the data of the network itself. This allows the utility to perceive and understand the scale and significance of illegal water use (and to note if any illegal use for irrigated agriculture is occurring in the network), and allows the utility to better manage and monitor unauthorised consumption. Likewise, after obtaining the total volume of unauthorised consumption in the network, this can be further broken down into subcomponents, in as much detail as possible, through the equations demonstrated in Chapter 9. This is the proposed methodology for assessing this challenging component. If the aforementioned methodology cannot be used because of the inability to apply the ALE equation suggested in Chapter 6, one would inevitably return to the starting point by estimating the unauthorised consumption through the default assumptions that affect all calculations of water loss assessment and reduction planning. To avoid this, the proposed method (and matrix) in Chapter 9 for estimating the unauthorised consumption based on the observed number of disconnected connections in the network is a practical method. This method remains, in all cases, more accurate than utilising abstract assumptions because it depends on vital and variable data coming from the network itself, not solely on data from other networks.

10.5   IMPROVED LEAKAGE ESTIMATION

If the accuracy of estimating the apparent losses is improved, the accuracy of the calculated real losses will necessarily improve. This is essential as many further necessary analyses depend mainly on the volume of the real losses in the network, and can then be carried out more reliably. Examples of these analyses include estimating the benefits of reducing network pressure or the economic feasibility of frequent active leakage detection surveys. After that, these analyses must be scaled down to the DMA scale, again to investigate the components of the apparent and real losses in DMAs and assess the potential of pressure management and other loss reduction options at the DMA scale. For this purpose, carrying out 'minimum flow' analysis in the DMAs (taking into account the aforementioned considerations) is inevitable and vital to arrive at a detailed water loss management strategy. Figures 10.2, 10.3, and 10.4 illustrate the sequential steps for assessing the water loss level and components in intermittent supply, beginning with the normalisation of the total water loss level, followed by the water loss assessment at the DMA scale, and ending with the determination of the economic level of water losses in the network. These flowcharts can be used as a starting point for more detailed guidance on the steps and methods of systematic water loss assessment in intermittent supply, which in turn contributes to more effective water loss management in intermittent supply networks, especially in developing countries.

10.6   THESIS CONTRIBUTION TO WATER LOSS MANAGEMENT

1. An extensive review and comparative analyses of existing methods and (software) tools for water loss assessment are provided (in this thesis) as well as their implication on the planning and management of water loss.

2. A new normalisation approach is proposed to monitor water loss management in individual intermittent supply networks as well as an approach to benchmark the performance to other networks with different supply time.

3. A novel method for the estimation of the overall volume of apparent losses in networks is proposed. In addition, alternative methodologies are introduced for the estimation of unauthorised water consumption as well as estimation and modelling customer meter inaccuracy in intermittently operated networks.

4. Guidance (and flow charts) for practical and improved leakage and apparent loss estimation is provided, which is a prerequisite to improve water loss management in intermittent water distribution networks.

10.7   FUTURE OUTLOOK

This study addressed many difficult problems in assessing water losses in intermittent supply networks. Nevertheless, many issues have emerged as requiring further investigation, the most important of which are the following:

(i)   Benchmarking water utilities with intermittent supply is a subject of increasing importance. This study suggested a normalisation approach to monitor the performance of water loss management by the utility using regression analysis. However, to compare and benchmark the water loss performance of different utilities, a standardisation of the w.s.p. normalisation is required at the national and international levels, and this needs to be further clarified. Benchmarking the apparent losses of different utilities using the w.s.p. adjustment is a thorny issue because it is susceptible to over-estimation.

(ii)   Regarding real losses, the water balance can be improved by following the methods presented in this study to estimate the real losses for the whole network. However, the problem of real loss assessment arises at the DMA level. The 'minimum flow' analysis can occasionally be applied but cannot be employed as a permanent solution, and there is a need to further investigate methods for DMA leakage estimation without disturbing the water distribution schedule. In addition, although modelling the pressure–leakage relationship is an issue that is being extensively researched, recent results have indicated the need to audit the laws of leakage modelling and clarify the pressure–leakage relationship, particularly the leakage exponent (N_1). Any progress in this area will have substantial repercussions, not only on the leakage assessment but also on the efficiency of the leakage detection technology, pressure management, and all other aspects of leakage reduction. Furthermore, it is very useful to review the parameters for analysing the leakage components using the BABE method, considering the specific conditions of intermittent supply. Revising the ICF may help bridge the gap between the output of this method and the actual situation in intermittent supply. Finally, in regard to planning the leakage reduction, estimating the potential benefits resulting from implementing leakage reduction interventions (e.g., pressure management) can be currently and independently analysed. The benefits of each option can be estimated separately, but this leads to an overestimation of the benefits of leakage reduction interventions because in reality these interventions influence each other. For example, reducing the network pressure reduces the number of bursts but limits the potential of leakage detection, as leakage becomes more challenging to detect with reduced pressure leaks in the network. Different leakage reduction interventions therefore affect each other, and ascertaining the interdependent relationships of these interventions is worth investigating.

(iii)   Considering apparent losses, the ALE equation proposed in this study is promising and was applied with acceptable results in several case studies in developing countries (Sana'a and Mwanza in this thesis, and two other networks in Jordan). However, there is still a need to test this method in developed countries and optimise its use by establishing typical network curves correlating billed consumption to the WWTP inflow. This will contribute to increasing the effectiveness of this method and further facilitating its application. In addition, this study proposed a matrix for the initial evaluation of unauthorised consumption by utilising the number of disconnected connections in the network. This matrix was developed based on data from six case studies, and additional case studies would inevitably enrich this matrix and aid in fine-tuning its parameters. Investigating the potential of artificial neural networks for the assessment and detection of unauthorised consumption of utility customers appears to be a promising research topic. In addition, this study clarified the effect of three different types and sizes of FVs on the water meter accuracy, but testing further types of FV characteristics and analysing their effect on the accuracy of the meters will further elucidate the impacts of FVs on the water meter performance. In this context, specific consumption data were used to model the water level in the tank. The intensity of consumption or the time period between the consumption events played an important role in determining the effect of the tank size on the accuracy of the water meter. Utilising other consumption data will therefore deepen our understanding of the effect of the tank size on the meter accuracy.

(iv) Finally, there is, importantly, a high potential for developing a comprehensive software tool that accommodates the aspects of water loss assessment, monitoring and management planning. Such a comprehensive tool has not yet been developed, however many smaller individual tools exist, addressing specific issues related to water loss management. Most of the currently available tools are not interconnected and lack a clear roadmap for developing a water loss management strategy. For this purpose, this study proposed guidance flowcharts for intermittent supply in Figures 10.2, 10.3, and 10.4. It is necessary for these processes to be reflected in a comprehensive water loss management tool (recognising network intermittency) so that the analyses within these tools are rendered sequential and closely related to one another, in order to reach a useful and practical final outcome that assists in formulating a water loss management strategy. In this regard, it should be noted that the scarcity of software tools for management of apparent losses is one of the manifestations of the insufficient attention to address the apparent losses to date. This problem should be addressed, as apparent losses are a central component of water losses in distribution networks in many countries worldwide, especially in the developing world.

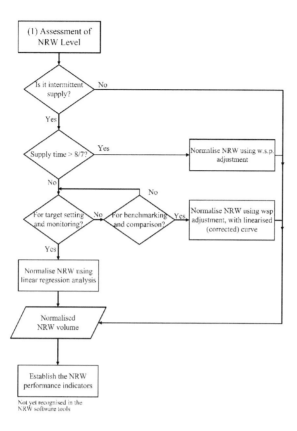

Figure 10.2. Flowchart for normalising the water loss level in intermittent supply networks

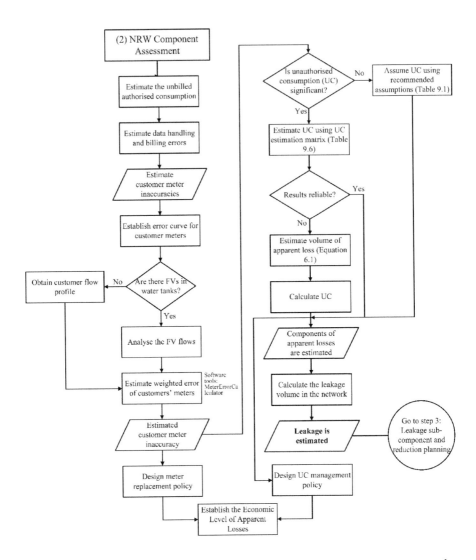

Figure 10.3. Flow chart for water loss component assessment in intermittent supply

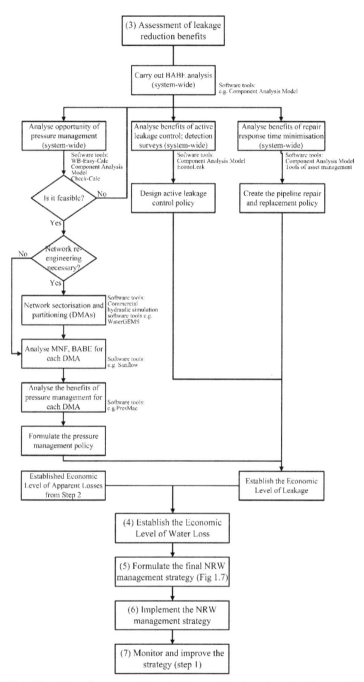

Figure 10.4. Flowchart for strategising leakage reduction planning (steps 4–7 are not covered in this study)

201

REFERENCES

A&N Technical Services Inc, Maureen Erbeznik & Associates, M.Cubed, and Associates, G. F. (2013). "Analysis of water use efficiency metrics and benchmarking." Mesa Water District, California, USA.

Aboelnga, H., Saidan, M., Al-Weshah, R., Sturm, M., Ribbe, L., and Frechen, F.-B. (2018). "Component analysis for optimal leakage management in Madaba, Jordan." *Journal of Water Supply: Research and Technology-Aqua*, 67 (4), 384-396.

Al-Ansari, N., Alibrahiem, N., Alsaman, M., and Knutsson, S. (2014). "Water Supply Network Losses in Jordan." *Journal of Water Resource and Protection*, 6(2), 83-96.

Al-Omari, A. (2013). "A Methodology for the Breakdown of NRW into Real and Administrative Losses." *Water resources management*, 27(7), 1913-1930.

Al-Washali, T. (2011). "Non-revenue water management in Sana'a water distribution system," Colgone University of Applied Sciences, Germany.

AL-Washali, T., Mahardani, M., Sharma, S., Arregui, F., and Kennedy, M. (2020a). "Impact of float-valves on customer meter performance under intermittent and continuous supply conditions." *Resources, Conservation and Recycling*, 163, 105091.

AL-Washali, T., Sharma, S., AL-Nozaily, F., Haidera, M., and Kennedy, M. (2019a). "Monitoring the Non-Revenue Water Performance in Intermittent Supplies." *Water*, 11(6), 1220.

AL-Washali, T., Sharma, S., and Kennedy, M. (2016). "Methods of Assessment of Water Losses in Water Supply Systems: a Review." *Water Resources Management*, 30(14), 4985-5001.

Al-Washali, T., Sharma, S., Lupoja, R., Al-Nozaily, F., Haidera, M., and Kennedy, M. (2020b). "Assessment of water losses in distribution networks: Methods, applications, uncertainties, and implications in intermittent supply." *Resources, Conservation and Recycling*, 152(1), 104515.

AL-Washali, T. M., Sharma, S. K., and Kennedy, M. D. (2018). "Alternative Method for Nonrevenue Water Component Assessment." *Journal of Water Resources Planning and Management*, 144(5), 04018017.

AL-Washali, T. M., Sharma, S. K., Kennedy, M. D., AL-Nozaily, F., and Mansour, H. (2019b). "Modelling the Leakage Rate and Reduction Using Minimum Night Flow Analysis in an Intermittent Supply System." *Water*, 11(1), 48.

Alegre, H., Baptista, J. M., Cabrera Jr, E., Cubillo, F., Duarte, P., Hirner, W., Merkel, W., and Parena, R. (2006). *Performance indicators for water supply services*, IWA publishing.

Alegre, H., Baptista, J. M., Cabrera Jr, E., Cubillo, F., Duarte, P., Hirner, W., Merkel, W., and Parena, R. (2016). *Performance indicators for water supply services*, IWA Publishing, London, UK.

Alegre, H., Hirner, W., and Baptista, J. (2000). *Performance Indicators for Water Supply Services. Manual of Best Practice*, IWA Publishing, London, UK.

Alkasseh, J. M., Adlan, M. N., Abustan, I., Aziz, H. A., and Hanif, A. B. M. (2013). "Applying minimum night flow to estimate water loss using statistical modeling: a case study in Kinta Valley, Malaysia." *Water Resources Management*, 27(5), 1439-1455.

Almandoz, J., Cabrera, E., Arregui, F., Cabrera Jr, E., and Cobacho, R. (2005). "Leakage assessment through water distribution network simulation." *Journal of water resources planning and management*, 131(6), 458-466.

Alonso, J. M., Alvarruiz, F., Guerrero, D., Hernández, V., Ruiz, P. A., Vidal, A. M., Martínez, F., Vercher, J., and Ulanicki, B. (2000). "Parallel computing in water network analysis and leakage minimization." *Journal of Water Resources Planning and Management*, 126(4), 251-260.

Amoatey, P., Minke, R., and Steinmetz, H. (2018). "Leakage estimation in developing country water networks based on water balance, minimum night flow and component analysis methods." *Water Practice and Technology*, 13(1), 96-105.

Aquadas-QS. (2007). *Aquadas QS—the Modular Software Solution for Water Works Management*, Vienna, Austria.

Araujo, L. S., Ramos, H., and Coelho, S. T. (2006). "Pressure Control for Leakage Minimisation in Water Distribution Systems Management." *Water Resources Management*, 20(1), 133-149.

Arbués, F., García-Valiñas, M. a. Á., and Martínez-Espiñeira, R. (2003). "Estimation of residential water demand: a state-of-the-art review." *The Journal of Socio-Economics*, 32(1), 81-102.

Arregui, F., Balaguer, M., Soriano, J., and García-Serra, J. (2015). "Quantifying measuring errors of new residential water meters considering different customer consumption patterns." *Urban Water Journal*, 13(5), 463-475.

Arregui, F., Cabrera, E., Cobacho, R., and García-Serra, J. (2005). "Key factors affecting water meter accuracy." IWA Leakage 2005 Conference, Halifax, Canada, International Water Association, London, UK.

Arregui, F., Cabrera Jr, E., and Cobacho, R. (2006a). *Integrated water meter management*, IWA Publishing, London, UK.

Arregui, F., Cabrera Jr, E., and Cobacho, R. (2007). *Integrated water meter management*, IWA Publishing, London, UK.

Arregui, F., Cobacho, R., Soriano, J., and Jimenez-Redal, R. (2018a). "Calculation Proposal for the Economic Level of Apparent Losses (ELAL) in a Water Supply System." *Water*, 10(12), 1809.

Arregui, F., Gavara, F., Soriano, J., and Pastor-Jabaloyes, L. (2018b). "Performance analysis of ageing single-jet water meters for measuring residential water consumption." *Water*, 10(5), 612.

Arregui, F. J., Cabrera, E., Cobacho, R., and García-Serra, J. (2006b). "Reducing Apparent Losses Caused By Meters Inaccuracies." *Water Practice and Technology*, 1(4), 2006093.

Ashton, C., and Hope, V. (2001). "Environmental valuation and the economic level of leakage." *Urban Water*, 3(4), 261-270.

Austin Water Utility. (2009). "Water loss audit report - Texas." Austin Water Utility, Texas, USA.

AWWA. (2009). *Water Audit and Loss Control Program*, American Water Works Association, Denver, USA.

AWWA. (2016). *M36 Water Audits and Loss Control Programs, Fourth Edition*, American Water Works Association, USA.

Buchberger, S. G., and Nadimpalli, G. (2004). "Leak estimation in water distribution systems by statistical analysis of flow readings." *Journal of Water Resources Planning and Management*, 130(4), 321-329.

Carpenter, T., Lambert, A., and McKenzie, R. (2003). "Applying the IWA approach to water loss performance indicators in Australia." *Water Science and Technology: Water Supply*, 3(1-2), 153-161.

Carteado, F., and Vermersch, M. (2010). "Apparent water losses generated by unauthorised consumption: assessment and corrective actions." International Water Association - Water Loss Specialist Group, London, UK.

Charalambous, B. (2019). "Ten Crucial Reasons to avoid Intermittent Water Supply." IWA 1st Intermittent Water Supply Conference, Kampala, Uganda, International Water Association, IWA Publishing, London, UK.

Charalambous, B., and Laspidou, C. (2017). *Dealing with the Complex Interrelation of Intermittent Supply and Water Losses*, IWA Publishing, London, UK.

Chisakuta, S., Mayumbelo, K., Mulenga, K., Simbeye, I., Wegelin, W., Mckenzie, R., Hamilton, S., and Anders, D. (2011). *Non-revenue water: trainers manual*, Wave Pool Zambia Imprint. German Foundation for International Cooperation (GIZ), Germany.

Christodoulou, S., Charalambous, C., and Adamou, A. (2008). "Rehabilitation and maintenance of water distribution network assets." *Water Supply*, 8(2), 231-237.

Claudio, K., Couallier, V., Leclerc, C., Le Gat, Y., and Saracco, J. (2015). "Consumption estimation with a partial automatic meter reading deployment." *Water Science and Technology: Water Supply*, 15(1), 50-58.

Colebrook, C., and White, C. (1937). "Experiments with fluid friction in roughened pipes." *Proceedings of the Royal Society of London. Series A-Mathematical and Physical Sciences*, 161(906), 367-381.

Colebrook, C. F. (1939). "Turbulent flow in pipes, with particular reference to the transition region between the smooth and rough pipe laws.(includes plates)." *Journal of the Institution of Civil Engineers*, 11(4), 133-156.

Couvelis, F., and Van Zyl, J. (2015). "Apparent losses due to domestic water meter under-registration in South Africa." *Water SA*, 41(5), 698-704.

Crane_Co. (1957). *Flow of fluids through valves, fittings, and pipe*, Crane Company - Engineering Division, Stamford, USA.

Creaco, E., and Pezzinga, G. (2014). "Multiobjective optimization of pipe replacements and control valve installations for leakage attenuation in water distribution networks." *Journal of Water Resources Planning and Management*, 141(3), 04014059.

Criminisi, A., Fontanazza, C., Freni, G., and Loggia, G. L. (2009). "Evaluation of the apparent losses caused by water meter under-registration in intermittent water supply." *Water Science and Technology*, 60(9), 2373-2382.

Cunha Marques, R., and Monteiro, A. (2003). "Application of performance indicators to control losses-results from the Portuguese water sector." *Water Science and Technology: Water Supply*, 3(1-2), 127-133.

Dai, P. D., and Li, P. (2014). "Optimal Localization of Pressure Reducing Valves in Water Distribution Systems by a Reformulation Approach." *Water Resources Management*, 28(10), 3057-3074.

Danilenko, A., Van den Berg, C., Macheve, B., and Moffitt, L. J. (2014). *The IBNET water supply and sanitation blue book 2014: The international benchmarking network for water and sanitation utilities databook*, The World Bank, Washington, USA.

De Marchis, M., Fontanazza, C., Freni, G., La Loggia, G., Notaro, V., and Puleo, V. (2013). "A mathematical model to evaluate apparent losses due to meter under-registration in intermittent water distribution networks." *Water Science and Technology: Water Supply*, 13(4), 914-923.

De Marchis, M., Fontanazza, C., Freni, G., Milici, B., and Puleo, V. (2014). "Experimental investigation for local tank inflow model." *Procedia Engineering*, 89, 656-663.

De Marchis, M., Fontanazza, C. M., Freni, G., Loggia, G. L., Napoli, E., and Notaro, V. (2010). "Analysis of the impact of intermittent distribution by modelling the network-filling process." *Journal of Hydroinformatics*, 13(3), 358-373.

De Marchis, M., Milici, B., and Freni, G. (2015). "Pressure-Discharge Law of Local Tanks Connected to a Water Distribution Network: Experimental and Mathematical Results." *Water*, 7(9), 4701-4723.

De Paola, F., Fontana, N., Galdiero, E., Giugni, M., Savic, D., and Sorgenti degli Uberti, G. (2014). "Automatic multi-objective sectorization of a water distribution network." *Procedia Engineering*, 89, 1200-1207.

Deuerlein, J. W. (2008). "Decomposition model of a general water supply network graph." *Journal of Hydraulic Engineering*, 134(6), 822-832.

Di Nardo, A., Di Natale, M., and Di Mauro, A. (2013a). *Water supply network district metering: theory and case study*, Springer Science & Business Media, Berlin, Germany.

Di Nardo, A., Di Natale, M., Gisonni, C., and Iervolino, M. (2015). "A genetic algorithm for demand pattern and leakage estimation in a water distribution network." *Journal of Water Supply: Research and Technology-Aqua*, 64(1), 35-46.

Di Nardo, A., Di Natale, M., Santonastaso, G. F., and Venticinque, S. (2013b). "An automated tool for smart water network partitioning." *Water Resources Management*, 27(13), 4493-4508.

Dighade, R., Kadu, M., and Pande, A. (2014). "Challenges in water loss management of water distribution systems in developing countries." *International Journal of Innovative Research in Science, Engineering and Technology*, 3(6), 13838-13846.

Dublin Drainage Consultancy. (2005). "Greater Dublin Strategic Drainage Study (Final Report)." Dublin, Ireland.

Ellis, B., and Bertrand-Krajewski, J.-L. (2010). *Assessing infiltration and exfiltration on the performance of urban sewer systems*, International Water Association, IWA Publishing, London, UK.

Ellis, J. B., Revitt, D. M., Lister, P., Willgress, C., and Buckley, A. (2003). "Experimental studies of sewer exfiltration." *Water Science and Technology*, 47(4), 61-67.

Ellis, J. B., Revitt, D. M., Vollertsen, J., and Blackwood, D. J. (2008). "Factors influencing temporal exfiltration rates in sewer systems." Proceedings of the 11th International Conference on Urban Drainage (11ICUD), International Water Association, IWA Publishing, London, UK, Edinburgh, Scotland.

EPA, U. S. E. P. A. (2010). *Control and mitigation of drinking water losses in distribution systems*, United States Environmental Protection Agency (US EPA) Washington DC, USA.

EPA, U. S. E. P. A. (2013). *Water audits and water loss control for public water systems*, Environmental Protection Agency, Washington DC, USA.

Erickson, J. J., Smith, C. D., Goodridge, A., and Nelson, K. L. (2017). "Water quality effects of intermittent water supply in Arraiján, Panama." *Water Research*, 114, 338-350.

Ermini, R., Ataoui, R., and Qeraxhiu, L. (2015). "Performance indicators for water supply management." *Water Science and Technology: Water Supply*, 15(4), 718-726.

Eugine, M. (2017). "Predictive Leakage Estimation using the Cumulative Minimum Night Flow Approach." *American Journal of Water Resources*, 5(1), 1-4.

European Commission, D. G. f. t. E. (2015). *EU Reference Document Good Practices on Leakage Management WFD CIS WG PoM: Main Report.*, European Union, Brussels.

Fanner, P. (2004). "Assessing real water losses: a practical approach." *Water 21*(April), 49-50.

Fanner, P., and Lambert, A. "Calculating SRELL with pressure management, active leakage control and leak run-time options, with confidence limits." *Proceedings of 5th IWA Water Loss Reduction Specialist Conference 2009*, Cape Town, South Africa, 373-380.

Fanner, P., and Thornton, J. "The importance of real loss component analysis for determining the correct intervention strategy." *Proceedings of IWA Water Loss 2005 Conference, Halifax, Nova Scotia, Canada*.

Fantozzi, M., and Lambert, A. "Legitimate night use component of minimum night flows initiative." *IWA Water Loss Conference 2010* São Paulo, Brazil.

Fantozzi, M., and Lambert, A. (2012). "Residential night consumption–assessment, choice of scaling units and calculation of variability." IWA Water Loss Conference 2012, Manila, Philippines, 26-29.

Farah, E., and Shahrour, I. (2017). "Leakage Detection Using Smart Water System: Combination of Water Balance and Automated Minimum Night Flow." *Water Resources Management*, 31(15), 4821-4833.

Farley, M., and Liemberger, R. (2005). "Developing a non-revenue water reduction strategy: planning and implementing the strategy." *Water Supply*, 5(1), 41-50.

Farley, M., and Trow, S. (2003). *Losses in water distribution networks*, IWA Publishing, London, UK.

Farley, M., Wyeth, G., Ghazali, Z. B. M., Istandar, A., Singh, S., Dijk, N., Raksakulthai, V., and Kirkwood, E. (2008). *The manager's non-revenue water handbook: a guide to understanding water losses*, Ranhill Utilities Berhad and the United States Agency for International Developement, Bangkok, Thailand.

Fontanazza, C. M., Notaro, V., Puleo, V., and Freni, G. (2015). "The apparent losses due to metering errors: a proactive approach to predict losses and schedule maintenance." *Urban Water Journal*, 12(3), 229-239.

Frauendorfer, R., and Liemberger, R. (2010). *The issues and challenges of reducing non-revenue water*, Asian Development Bank, Metro Manila, Philippines.

Galdiero, E., De Paola, F., Fontana, N., Giugni, M., and Savic, D. (2015). "Decision support system for the optimal design of district metered areas." *Journal of Hydroinformatics*, 18(1), 49-61.

Galdiero, E., De Paola, F., Fontana, N., Giugni, M., and Savic, D. (2016). "Decision support system for the optimal design of district metered areas." *Journal of Hydroinformatics*, 18(1), 49-61.

Giustolisi, O., Savic, D., and Kapelan, Z. (2008). "Pressure-driven demand and leakage simulation for water distribution networks." *Journal of Hydraulic Engineering*, 134(5), 626-635.

Gleick, P., Haasz, D., Henges-Jeck, C., Srinivasan, V., Wolff, G., Cushing, K. K., and Mann, A. (2003). *Waste not, want not: The potential for urban water conservation in California*, Pacific Institute for Studies in Development, Environment, and Security Oakland, CA, USA.

Gong, W., Suresh, M. A., Smith, L., Ostfeld, A., Stoleru, R., Rasekh, A., and Banks, M. K. (2016). "Mobile sensor networks for optimal leak and backflow detection and localization in municipal water networks." *Environmental Modelling & Software*, 80, 306-321.

Greyvenstein, B., and van Zyl, J. E. (2007). "An experimental investigation into the pressure - leakage relationship of some failed water pipes." *Journal of Water Supply: Research and Technology-Aqua*, 56(2), 117-124.

GWR-Ltd. (2008). *Water Losses Performance Indicators Program "BenchLoss" for New Zeland*, Global Water Resources Ltd, New Zealand.

Halfawy, M. R., and Hunaidi, O. (2008). "GIS-based water balance system for integrated sustainability management of water distribution assets." 60th Annual Western Canada Water and Wastewater Association Conference, Regina, Saskatchewan, Canada, pp. 1-16.

Hamilton, S., and McKenzie, R. (2014). *Water management and water loss*, IWA Publishing, London, UK.

Herrera, M., Izquierdo, J., Pérez-García, R., and Ayala-Cabrera, D. (2010). "Water supply clusters by multi-agent based approach." 12th Annual Conference on Water Distribution Systems Analysis 2010, Tucson, Arizona, United States, 861-869.

Heydenreich, M., and Kreft, D. (2004). "Factor of Success Network Management." *energie |wasser-praxis*, 55(12:84).

Høgh, K. "Assessment of Real Losses from Minimum Night Flows." *Proceedings of IWA Specialized Conference: Water Loss 2014*, Austria, Vienna.

IBNET. (2020). "IB-NET Database. Accessed on 23/07/2020 from: https://database.ib-net.org/DefaultNew.aspx. International Benchmarking Network."

ISO. (2014a). "Water Meters for Cold Potable Water and Hot Water—Part 1: Metrological and Technical Requirements." ISO 4064-1, International Organization for Standardization, Geneva, Switzerland.

ISO. (2014b). "Water Meters for Cold Potable Water and Hot WaterWater—Part 3: Test Report Format." ISO 4064-3, International Organization for Standardization, Geneva, Switzerland.

ITA. (2000). "SigmaLite Software: IWAP Manual of Best Practice and PIs for Water Supply Services." Instituto Tecnológico del Agua València, Spain.

Jernigan, W. (2014). "Next Generation Water Loss Tools - AWWA's New and Improving Tools and Publications for Water Loss Control." North Carolina, USA.

JICA. (2007). "The Study for the Water Resources Management and Rural Water Supply Improvement in the Republic of Yemen, Water Resources Management Action Plan for Sana'a Basin, Final Report." Japan International Cooperation Agency, Tokyo, Japan.

Johnson, E. (2001). "Optimal water meter selection system." *Water SA*, 27(4), 481-488.

Jowitt, P. W., and Xu, C. (1990). "Optimal valve control in water-distribution networks." *Journal of Water Resources Planning and Management*, 116(4), 455-472.

Kanakoudis, V., and Tsitsifli, S. (2014). "Using the bimonthly water balance of a non-fully monitored water distribution network with seasonal water demand peaks to define its actual NRW level: the case of Kos town, Greece." *Urban Water Journal*, 11(5), 348-360.

Kanakoudis, V., Tsitsifli, S., and Papadopoulou, A. (2012). "Integrating the carbon and water footprints' costs in the water framework directive 2000/60/EC full water cost recovery concept: basic principles towards their reliable calculation and socially just allocation." *Water*, 4(1), 45-62.

Kesavan, H., and Chandrashekar, M. (1972). "Graph-theoretical models for pipe network analysis." *Journal of the Hydraulics Division*, 98(2), 345-364.

KFW. (2008). " Water loss reduction programme: evaluation report. German Bank for Development (Kreditanstalt für Wiederaufbau)." Frankfurt, Germany.

Kingdom, B., Liemberger, R., and Marin, P. (2006). *The challenge of reducing non-revenue water (NRW) in developing countries - paper 8*, The World Bank Washington DC, USA.

Klingel, P., and Knobloch, A. (2015). "A review of water balance application in water supply." *Journal-American Water Works Association*, 107(7), E339-E350.

Koldžo, Ð., and Vuc̆ijak, B. (2013). "Testing innovative software tool CalcuLEAKator for water balance evaluation and water loss reduction in Tuzla project." 6th International Water Loss Conference, Sofia, Bulgaria.

Korkmaz, N., and Avci, M. (2012). "Evaluation of water delivery and irrigation performances at field level: The case of the menemen left bank irrigation district in Turkey." *Indian Journal of Science and Technology*, 5(2), 2079-2089.

Lambert, A. (1994). "Accounting for losses: The bursts and background concept." *Water and Environment Journal*, 8(2), 205-214.

Lambert, A. "Pressure management/leakage relationships: Theory, concepts and practical applications." *Conference on Minimising Losses in Water Supply Systems 1997, IQPC Ltd*, London, UK.

Lambert, A. "What do we know about pressure-leakage relationships in distribution systems." *IWA Conference in Systems approach to leakage control and water distribution system management 2001*, Brno, Czech Republic.

Lambert, A. (2002). "International report: water losses management and techniques." *Water Science and Technology: Water Supply*, 2(4), 1-20.

Lambert, A. (2003). "Assessing non-revenue water and its components: a practical approach." Water 21, IWA, London, UK, 50-51.

Lambert, A. (2009). "Ten years experience in using the UARL formula to calculate infrastructure leakage index." Proceedings of IWA Specialized Conference:

Water Loss 2009, International Water Association, UK, Cape Town, South Africa, 189-196.

Lambert, A. (2015a). "CheckCalcs: Free Check on Leakage and Pressure Management Opportunities (v6b)." ILMSS Ltd – International Leakage Management Support Services Ltd, UK.

Lambert, A. (2015b). "LEAKS: Leakage Evaluation and Assessment Know-How Software." ILMSS Ltd – International Leakage Management Support Services Ltd, UK.

Lambert, A. (2018). "Fast Track NDF calculations using the Correction Factor method."

Lambert, A. (2019). "Interpreting Leakage in England & Wales." Leaks Suite Library, UK.

Lambert, A., Brown, T. G., Takizawa, M., and Weimer, D. (1999). "A review of performance indicators for real losses from water supply systems." *Journal of Water Supply: Research and Technology-AQUA*, 48(6), 227-237.

Lambert, A., Charalambous, B., Fantozzi, M., Kovac, J., Rizzo, A., and St John, S. G. (2014). "14 years' experience of using IWA best practice water balance and water loss performance indicators in Europe." Proceedings of IWA Specialized Conference: Water Loss 2014, Vienna, Austria, International Water Association, London, UK.

Lambert, A., and Fantozzi, M. (2005). "Recent advances in calculating economic intervention frequency for active leakage control, and implications for calculation of economic leakage levels." *Water Science and Technology: Water Supply*, 5(6), 263-271.

Lambert, A., Fantozzi, M., and Shepherd, M. (2017a). "FAVAD Pressure & Leakage:How Does Pressure Influence N1?" IWA Water Efficient 2017 Conference, 18 -20 July 2017, Bath, UK., International Water Association, London, UK, Bath, UK.

Lambert, A., Fantozzi, M., and Shepherd, M. (2017b). "Pressure: Leak flow rates using FAVAD: An improved fast-track practitioner's approach." Proceedings of the Computing and Control for the Water Industry (CCWI 2017), Sheffield, UK.

Lambert, A., and Hirner, W. (2000). *Losses from water supply systems: a standard terminology and recommended performance measures*, IWA Publishing, London, UK.

Lambert, A., and Lalonde, A. (2005). "Using practical predictions of economic intervention frequency to calculate short-run economic leakage level, with or without pressure management." Proceedings of IWA Specialised Conference - Leakage 2005 International Water Association, Uk, Halifax, Nova Scotia, Canada, 310-321.

Lambert, A., and McKenzie, R. "Practical experience in using the Infrastructure Leakage Index." *Proceedings of IWA Conference–Leakage Management: A Practical Approach* Lemesos, Cyprus.

Lambert, A., and Morrison, J. (1996). "Recent developments in application of 'bursts and background estimates' concepts for leakage management." *Water and Environment Journal*, 10(2), 100-104.

Lambert, A., and Taylor, R. (2010). "Water loss guidelines." Water New Zealand, Wellington, New Zealand.

Latchoomun, L., King, R. A., and Busawon, K. (2015). "A new approach to model development of water distribution networks with high leakage and burst rates." *Procedia Engineering*, 119, 690-699.

Laucelli, D., and Meniconi, S. (2015). "Water distribution network analysis accounting for different background leakage models." *Procedia Engineering*, 119, 680-689.

Li, R., Huang, H., Xin, K., and Tao, T. (2014). "A review of methods for burst/leakage detection and location in water distribution systems." *Water Supply*, 15(3), 429-441.

Liemberger, and Partners. (2018). "WB-EasyCalc software (v. 5.16)." Liemberger & Partners GmbH, Pressbaum, Austria.

Liemberger, R. "Recommendations for Initial Non-Revenue Water Assessment." *IWA Water Loss 2010*, São Paulo, Brazil.

Liemberger, R., and Farley, M. (2004). "Developing a nonrevenue water reduction strategy Part 1: Investigating and assessing water losses." Proceedings of IWA Specialized Conference: the 4th IWA World Water Congress International Water Association, UK, Marrakech, Morocco.

Liemberger, R., and McKenzie, R. "Aqualibre: a new innovative water balance software." *IWA and AWWA Conference on Efficient Management of Urban Water Supply*, Tenerife, Spain.

Liemberger, R., and Wyatt, A. (2018). "Quantifying the global non-revenue water problem." *Water Science and Technology: Water Supply*, 19(3), 831-837.

Male, J. W., Noss, R. R., and Moore, I. C. (1985). *Identifying and reducing losses in water distribution systems*, Noyes Publications, New Jersey, USA.

Mantilla-Peña, C. F., Widdowson, M. A., and Boardman, G. D. (2018). "Evaluation of in-service residential nutating disc water meter performance." *AWWA Water Science*, 1(1), e1113.

May, J. (1994). "Pressure Dependent Leakage." World Water and Environmental Engineering, Water Environment Federation, Washington DC.

Mayer, P. W., DeOreo, W. B., Opitz, E. M., Kiefer, J. C., Davis, W. Y., Dziegielewski, B., and Nelson, J. O. (1999). "Residential end uses of water." American Water Works Association, Denver, USA.

Mazzolani, G., Berardi, L., Laucelli, D., Martino, R., Simone, A., and Giustolisi, O. (2016). "A methodology to estimate leakages in water distribution networks based on inlet flow data analysis." *Procedia Engineering*, 162, 411-418.

Mazzolani, G., Berardi, L., Laucelli, D., Simone, A., Martino, R., and Giustolisi, O. (2017). "Estimating Leakages in Water Distribution Networks Based Only on Inlet Flow Data." *Journal of Water Resources Planning and Management*, 143(6), 04017014.

McIntosh, A. C. (2003). *Asian water supplies reaching the urban poor*, Asian Development Bank, Metro Manila, Philippines.

Mckenzie, R. (1999). *Development of a standardised approach to evaluate burst and background losses in potable water distribution systems : SANFLOW User Guide*, South African Water Research Commission, Pretoria, South Africa.

Mckenzie, R. (2007). *AquaLite Water Balance Software*, South Africa Water Research Commission, Pretoria, South Africa.

McKenzie, R., Buckle, H., Wegelin, W., and Meyer, N. (2003). *Water Demand Management Cookbook*, Rand Water, Johannesburg, South Africa.

Mckenzie, R., and Lambert, A. (2002). *Development of a Windows based package for assessing appropriate levels of active leakage control in potable water distribution systems - ECONOLEAK User Guide*, South African Water Research Commission, Pretoria, South Africa.

McKenzie, R., and Lambert, A. (2004). "Best practice performance indicators: a practical approach." *Water 21*, 43-45.

Mckenzie, R., Lambert, A., Kock, J., and Mtshweni, W. (2002). *Development of a simple and pragmatic approach to benchmark real losses in potable water distribution systems - BENCHLEAK User Guide*, South African Water Research Commission, Pretoria, South Africa.

Mckenzie, R., and Langenhoven, S. (2001). *Development of a pragmatic approach to evaluate the potential savings from pressure management in potable water distribution systems - PRESMAC User Guide*, South African Water Research Commission, Pretoria, South Africa.

MDWSD. (2011). "Annual water loss reduction plan implementation status 2010." Miami, USA.

Mekonnen, M. M., and Hoekstra, A. Y. (2016). "Four billion people facing severe water scarcity." *Science advances*, 2(2), e1500323.

Meseguer, J., Mirats-Tur, J. M., Cembrano, G., Puig, V., Quevedo, J., Pérez, R., Sanz, G., and Ibarra, D. (2014). "A decision support system for on-line leakage localization." *Environmental Modelling & Software*, 60, 331-345.

Mimi, Z., Abuhalaweh, O., Wakileh, V., and Staff, J. W. U. (2004). "Evaluation of water losses in distribution networks: Rammallah as a case study." *Water Science and Technology: Water Supply*, 4(3), 183-195.

Mini, C., Hogue, T., and Pincetl, S. (2014). "Estimation of residential outdoor water use in Los Angeles, California." *Landscape and Urban Planning*, 127(July), 124-135.

Mitchell, G., Mein, R., and McMahon, T. (1999). *The reuse potential of urban stormwater and wastewater*, Cooperative Research Centre for Catchment Hydrology, Victoria, Australia.

Moahloli, A., Marnewick, A., and Pretorius, J. (2019). "Domestic water meter optimal replacement period to minimize water revenue loss." *Water SA*, 45(2), 165-173.

Molinos-Senante, M., Mocholí-Arce, M., and Sala-Garrido, R. (2016). "Estimating the environmental and resource costs of leakage in water distribution systems: A shadow price approach." *Science of The Total Environment*, 568, 180-188.

Morrison, J., Tooms, S., and Rogers, D. (2007). "District Metered Areas, Guidance Notes." International Water Association (IWA), Specialist Group on Efficient Operation and Management of Urban Water Distribution Systems, UK.

Mutikanga, H. (2012). "Water loss management: tools and methods for developing countries, PhD thesis," IHE-Delft, Institute for Water Education, Delft, The Netherlands.

Mutikanga, H., Sharma, S., and Vairavamoorthy, K. (2011a). "Assessment of apparent losses in urban water systems." *Water and Environment Journal*, 25(3), 327-335.

Mutikanga, H., Sharma, S., Vairavamoorthy, K., and Cabrera, E. (2010). "Using performance indicators as a water loss management tool in developing countries." *Journal of Water Supply: Research and Technology-AQUA*, 59(8), 471-481.

Mutikanga, H. E. (2014). "Residential water meter selection using the analytical hierarchy process." *Journal AWWA*, 106(5), E233-E241.

Mutikanga, H. E., Sharma, S. K., and Vairavamoorthy, K. (2011b). "Investigating water meter performance in developing countries: A case study of Kampala, Uganda." *Water SA*, 37(4), 567-574.

Mutikanga, H. E., Sharma, S. K., and Vairavamoorthy, K. (2012). "Methods and tools for managing losses in water distribution systems." *Journal of Water Resources Planning and Management*, 139(2), 166-174.

MWE. (2007). "Performance Indicators Information System 2007." Ministry of Water and Environment, Sana'a, Yemen.

Ncube, M., and Taigbenu, A. (2019). "Assessment of apparent losses due to meter inaccuracy–a comparative approach." *Water SA*, 45(2), 174-182.

Netbase. (2019). "Netbase software." Crowder Consulting, Merseyside, UK.

OIML. (2013a). "Water Meters Intended for the Metering of Cold Potable Water and Hot Water—Part 1: Metrological and Technical Requirments." OIML 49-1, International Organization of Legal Metrology, Paris, France.

OIML. (2013b). "Water Meters Intended for the Metering of Cold Potable Water and Hot Water—Part 2: Test Methods." OIML 49-2 International Organization of Legal Metrology, Paris, France.

Orden_ICT_155. (2020). "Orden ICT_155_2020 por la que se regula el control metrológico del Estado de determinados instrumentos de medida." BOE-A-2020-2573, Ministry of Industry, Commerce and Tourism, Spain.

ÖVGW. (2009). "ÖVGW Guideline W 63. Water Losses in Drinking Water Supply Systems. Austrian Standards." Vienna, Austria.

Palau, C., Arregui, F., and Carlos, M. (2012). "Burst detection in water networks using principal component analysis." *Journal of Water Resources Planning and Management*, 138(1), 47-54.

Palenchar, J., Friedman, K., and Heaney, J. P. (2009). "Hydrograph separation of indoor and outdoor billed water use in Florida's single family residential sector." Proceedings of the 2009 FS AWWA Fall Conference, American Water Works Association, Denver, USA, Orlando, USA.

Park, H. J. (2007). "A study to develop strategies for proactive water-loss management," Georgia Institute of Technology, Atlanta, GA, USA.

Pearson, D., and Trow, S. (2005). "Calculating economic levels of leakage." IWA Water Loss 2005 Conference, International Water Association, Uk, Halifax, NS, Canada.

Perelman, L., and Ostfeld, A. (2011). "Topological clustering for water distribution systems analysis." *Environmental Modelling & Software*, 26(7), 969-972.

Pillot, J., Catel, L., Renaud, E., Augeard, B., and Roux, P. (2016). "Up to what point is loss reduction environmentally friendly? The LCA of loss reduction scenarios in drinking water networks." *Water Research*, 104, 231-241.

Pillot, J., Renaud, E., and Clauzier, M. "A method of analysing night consumption in DMAs with high levels of seasonal variation." *Proceedings of IWA Specialized Conference: Water Loss 2014*, Austria, Vienna.

Puust, R., Kapelan, Z., Savic, D., and Koppel, T. (2010). "A review of methods for leakage management in pipe networks." *Urban Water Journal*, 7(1), 25-45.

Pybus, P., and Schoeman, G. (2001). "Performance indicators in water and sanitation for developing areas." *Water science and technology*, 44(6), 127-134.

Radivojević, D., Milićević, D., and Blagojević, B. (2008). "IWA best practice and performance indicators for water utilities in Serbia: Case study Pirot." *Facta universitatis-series: Architecture and Civil Engineering*, 6(1), 37-50.

Renaud, E., Clauzier, M., Sandraz, A., Pillot, J., and Gilbert, D. (2014). "Introducing pressure and number of connections into water loss indicators for French drinking water supply networks." *Water Science and Technology: Water Supply*, 14(6), 1105-1111.

Richards, G. L., Johnson, M. C., and Barfuss, S. L. (2010). "Apparent losses caused by water meter inaccuracies at ultralow flows." *Journal AWWA*, 102(5), 123-132.

Rizzo, A., and Cilia, J. (2005). "Quantifying meter under-registration caused by the ball valves of roof tanks (for indirect plumbing systems)." Proceedings of the IWA Leakage 2005 Conference, Halifax, Canada, International Water Association, London, UK.

Romero, C. C., and Dukes, M. D. (2010). "Residential benchmarks for minimal landscape water use." UF Water Institute, University of Florida, USA.

Rubinstein, R. Y., and Kroese, D. P. (2016). *Simulation and the Monte Carlo method*, John Wiley & Sons, New Jersey, USA.

Rutsch, M. (2006). "Assessment of sewer leakage by means of exfiltration measurements and modelling tests, PhD Thesis," Geo- und Hydrowissenschaften der Technischen Universität Dresden, Germany.

Rutsch, M., Rieckermann, J., Cullmann, J., Ellis, J. B., Vollertsen, J., and Krebs, P. (2008). "Towards a better understanding of sewer exfiltration." *Water Research*, 42(10), 2385-2394.

Salim, H., and Manurung, N. "Bottom up water balance estimation: applications and limitations." *Proceedings of IWA Specialized Conference: Water Loss 2012*, Ferrara, Italy.

Schwaller, J., and van Zyl, J. E. (2015). "Modeling the Pressure-Leakage Response of Water Distribution Systems Based on Individual Leak Behavior." *Journal of Hydraulic Engineering*, 141(5), 04014089.

Schwaller, J., van Zyl, J. E., and Kabaasha, A. M. (2015). "Characterising the pressure-leakage response of pipe networks using the FAVAD equation." *Water Supply*, 15(6), 1373-1382.

Seago, C., Bhagwan, J., and McKenzie, R. (2004). "Benchmarking leakage from water reticulation systems in South Africa." *Water SA*, 30(5), 25-32.

Shields, D. J., Barfuss, S. L., and Johnson, M. C. (2012). "Revenue recovery through meter replacement." *Journal AWWA*, 104(4), E252-E259.

Shiklomanov, I. (1993). "World fresh water resources." Water in crisis: A guide to the world's fresh water resources, P. Gleick, ed., Oxford University Press.

Singh, R., Maheshwari, B., and Malano, H. (2009). "Developing a conceptual model for water accounting in peri-urban catchments." Proceedings of the 18th World IMACS / MODSIM Congress, The Government of Queensland, Australia, Cairns, Australia, 13-17.

Sommerfeld, A. (1908). "Ein beitrag zur hydrodynamischen erklaerung der turbulenten fluessigkeitsbewegungen." The 4th International Congress of Mathematics, International Congress of Mathematicians, Rome, Italy, 116–124.

Ssozi, E. N., Reddy, B. D., and Zyl, J. E. v. (2016). "Numerical Investigation of the Influence of Viscoelastic Deformation on the Pressure-Leakage Behavior of Plastic Pipes." *Journal of Hydraulic Engineering*, 142(3), 04015057.

Staben, N., Hein, A., and Kluge, T. (2010). "Measuring sustainability of water supply: performance indicators and their application in a corporate responsibility report." *Water Science and Technology: Water Supply*, 10(5), 824-830.

Stoker, D. M., Barfuss, S. L., and Johnson, M. C. (2012). "Flow measurement accuracies of in-service residential water meters." *Journal AWWA*, 104(12), E637-E642.

Sturm, R., Gasner, K., Wilson, T., Preston, S., and Dickinson, M. A. (2014). *Real Loss Component Analysis: A Tool for Economic Water Loss Control*, Water Research Foundation, Denver, USA.

SWSLC. (2010). "Performance Indicators Information System 2010." Sana'a Water Supply and Sanitation Local Corporation, Sana'a, Yemen.

Tabesh, M., Yekta, A. H. A., and Burrows, R. (2009). "An Integrated Model to Evaluate Losses in Water Distribution Systems." *Water Resources Management*, 23(3), 477-492.

Tamari, S., and Ploquet, J. (2012). "Determination of leakage inside buildings with a roof tank." *Urban Water Journal*, 9(5), 287-303.

Taylor, J. (1997). *Introduction to error analysis, the study of uncertainties in physical measurements*, University Science Books, New York, USA.

Thornton, J., and Lambert, A. "Progress in practical prediction of pressure: leakage, pressure: burst frequency and pressure: consumption relationships." *Proceedings of IWA Specialised Conference - Leakage 2005*, Halifax, Nova Scotia, Canada, 12-14.

Thornton, J., and Rizzo, A. "Apparent losses, how low can you go?" *IWA leakage management conference*, Lemesos, Cyprus, 20-22.

Thornton, J., Sturm, R., and Kunkel, G. (2008). *Water Loss Control*, McGraw Hill Professional, New York, USA.

Tsitsifli, S., and Kanakoudis, V. "Presenting a new user friendly tool to assess the performance level & calculate the water balance of water networks." *PRE10 International Conference*, Lefkada Island, Greece.

UN. (2015). "Transforming our world: the 2030 Agenda for Sustainable Development: Resolution adopted by the General Assembly on 25 September 2015." U. G. A. Resolution, ed., United Nations, New York, USA.

UNHSP. (2012). *Reduction of Illegal Water*, United Nations Human Settlements Programme - UNHSP, Nairobi, Kenya.

Van den Berg, C., and Danilenko, A. (2010). *The IBNET Water Supply and Sanitation Performance Blue Book: The International Benchmarking Network for Water and Sanitation Utilities Databook*, World Bank Publications, USA.

Van Zyl, J., and Cassa, A. (2014). "Modeling elastically deforming leaks in water distribution pipes." *Journal of Hydraulic Engineering*, 140(2), 182-189.

Van Zyl, J., Lambert, A., and Collins, R. (2017). "Realistic Modeling of Leakage and Intrusion Flows through Leak Openings in Pipes." *Journal of Hydraulic Engineering*, 143(9), 04017030.

Van Zyl, J. E. (2011). *Introduction to integrated water meter management*, Water Research Commission, South Africa.

Vermersch, M., Carteado, F., Rizzo, A., Johnson, E., Arregui, F., and Lambert, A. (2016). "Guidance notes on apparent losses and water loss reduction planning." LeaksSuit Library, UK.

Vermersch, M., and Rizzo, A. (2008). "Designing an action plan to control non-revenue water." *Water*, 21, 39-41.

Vicente, D. J., Garrote, L., Sánchez, R., and Santillán, D. (2016). "Pressure Management in Water Distribution Systems: Current Status, Proposals, and Future Trends." *Journal of Water Resources Planning and Management*, 142(2), 04015061.

Wallace, L. P. (1987). *Water and revenue losses: unaccounted-for water*, American Water Works Association, Denver, CO, USA.

Walter, D., Mastaller, M., and Klingel, P. (2018). "Accuracy of single-jet and multi-jet water meters under the influence of the filling process in intermittently operated pipe networks." *Water Supply*, 18(2), 679-687.

Water Loss Control Committee. (2014). *AWWA Free Water Audit Software (version 5.0)*, Water Loss Control Committee, American Water Works Association, Denver, CO, USA.

WEF. (2019). "The global risks report 2019." Geneva, Switzerland.

Werner, M., Maggs, I., and Petkovic, M. "Accurate measurements of minimum night flows for water loss analysis." *5th Annual WIOA NSW Water Industry Engineers & Operators Conference. Water Loss Management Program NSW*, Newcastle, UK, 31-37.

WHO. (2020). "Water, sanitation and hygiene." World Health Organization, Geneva, Switzerland.

WHO/UNICEF. (2017). "Joint Monitoring Program for Water, Sanitation and Hygiene." World Health Organization, Geneva, Switzerland.

Wu, Y., and Liu, S. (2017). "A review of data-driven approaches for burst detection in water distribution systems." *Urban Water Journal*, 14(9), 972-983.

Wu, Z. Y., Farley, M., Turtle, D., Kapelan, Z., Boxall, J., Mounce, S., Dahasahasra, S., Mulay, M., and Kleiner, Y. (2011). *Water loss reduction*, Bentley Institute Press.

Wu, Z. Y., Sage, P., and Turtle, D. (2009). "Pressure-dependent leak detection model and its application to a district water system." *Journal of Water Resources Planning and Management*, 136(1), 116-128.

Wyatt, A. S. (2010). *Non-revenue water: financial model for optimal management in developing countries*, RTI Press No. MR-0018-1006 Durham, USA.

Yazdandoost, F., and Izadi, A. (2018). "An asset management approach to optimize water meter replacement." *Environmental Modelling & Software*, 104, 270-281.

Zallom, S. (2014). "Analysis of bills' complaints and their causes in Zarqa water supply system. Zarqa water utility, Jordan."

APPENDIX A1: APPARENT LOSS BREAKDOWN IN ZARQA WATER NETWORK

A1.1 INTRODUCTION

This appendix presents additional information on estimating the different components of apparent losses in the Zarqa water network, including the unbilled authorised consumption, customer meter inaccuracies, data handling errors, and billing errors. Typically, these components comprise revenue losses; however, in the Zarqa water network, some apparent losses are not revenue losses, but represent overbilling profits. This section examines the overbilling practice in the Zarqa water network and analyses the other components of apparent losses in the network.

A1.2 UNBILLED AUTHORISED CONSUMPTION (UAC) IN ZARQA

More than 1% of the SIV of the Zarqa water network (Jordan) is 'free water' for customers in the Azraq Oasis, which amounts to 180 m³/quarter/customer (extremely high volume for domestic consumption in Zarqa). However, this quantity of water is billed and therefore should not be considered in the volume of the UAC in Zarqa. In this analysis, the UAC was inventoried for 2014, and data gaps were filled with the available data from the years 2008 and 2009. The results showed that the UAC in Zarqa is 1,380,335 m³/year, which is 2.1% of the SIV. This quantity seems to be higher than what is expected from utilities of developed and developing countries, which is normally from 0.5–1.5% of the billed consumption (AL-Washali et al. 2018). Figure A.1 and Table A.1 show the components of the UAC in Zarqa.

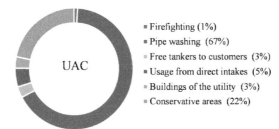

- Firefighting (1%)
- Pipe washing (67%)
- Free tankers to customers (3%)
- Usage from direct intakes (5%)
- Buildings of the utility (3%)
- Conservative areas (22%)

Figure A.1. UAC in Zarqa water network

Table A.1. Volume of UAC in Zarqa water network

	Q1		Q2		Q3		Q4		Total
	Metered	Unmetered	Metered	Unmetered	Metered	Unmetered	Metered	Unmetered	
Firefighting	1,126	1,000	1,126	1,000	1,126	1,000	1,126	1,000	8,504
Pipe washing		230,900		230,900		230,900		230,900	923,600
Free tankers to customers	7,609	1,388	7,609	952	7,609	1,668	7,609	1,708	36,152
Usage from direct intakes*	25,171		10,543		11,810		25,489		73,013
Buildings of the utility	12,022		9,248		9,248		9,248		39,766
Conservative areas	74,825		74,825		74,825		74,825		299,300
Total	120,753	233,288	103,351	232,852	104,618	233,568	118,297	233,608	1,380,335
								Total metered	447,019
								Total unmetered	933,316
								Grand Total	**1,380,335**
								%SIV	**2.1**

* Mainly from wells in the desert for Bedouin people and for livestock

A1.3   CUSTOMER METER INACCURACY

In Zarqa, households typically install one or several water tanks, either on the roof or on the ground, to store water in case the supply is interrupted. The tanks are usually made of a plastic material and commonly have capacities of 1 or 2 m³. Billed consumption passes through the float valve to fill the tanks. A study was conducted by the Zarqa water utility (Miyahuna) to assess the customer meter accuracy for different tank heights. Figure A.2 shows the flow rates of the float valve and the corresponding meter accuracy for a representative sample (assumed). To reflect the results of this experiment on the consumption data of the Zarqa water utility, the common height (H) of customer water tanks is required. Observations show that typical heights of water tanks in Zarqa are 1.18, 1.38, and 2 m for tanks with a 1 m³ capacity and 1.51 and 2.1 m for tanks with a 2 m³ capacity. The weighted average of customer meter under-registration was calculated as 3.8%, as shown in Table A.2.

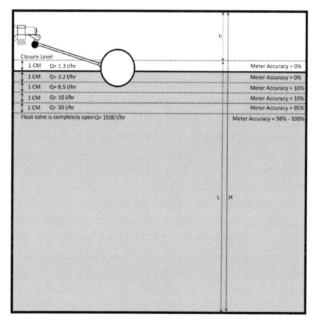

Figure A.2. Customer meter accuracy at different heights in the tank. Source of data: Zarqa water utility–Miyahuna

Table A.2. Customer meter inaccuracy in the Zarqa water network

Tank Capacity	H	h	L	Weighted Accuracy	Average	Prevalence	Meter Accuracy	Meter Inaccuracy
(m3)	(m)	(m)	(m)	(%)	(%)	(%)	(%)	(%)
	1.18	0.2	0.98	95%				
1.00	1.38	0.2	1.18	96%	96%	40%		
	2.00	0.2	1.80	97%			96.3%	3.7%
2.00	1.51	0.2	1.31	96%	97%	60%		
	2.10	0.2	1.90	97%				

A1.4   DATA HANDLING ERRORS

The changes to the actual readings of customer meters are data handling errors. Some errors are positive, but others are negative. Such errors increase or decrease utility revenues. The Zarqa water distribution network is divided into 49 reading areas with a meter reader for each area. The readings of the meters are collected quarterly (four times per year). A reader typically reads data from approximately 150 meters each working day. With an average of 3000 customers per area, a reader surveys a reading area within 20–25 days.

Since 2011, the Zarqa water utility has adopted the policy of registering the meter readings directly to hand-held units (hhu). These units can generate a water bill in the field. The data of the hhu is transferred wirelessly to the billing system through another software program in the server. Before a meter reader goes on his route, the hhu is programmed with data that includes the serial number of each customer, serial number of contracts, previous reading of the meters, and estimated average of historical consumption of eight previous cycles. If there is any mistake during the reading, such as reading the digit 6 as 9 or 6 as 8, the hhu does not approve the reading but provides an alarm. If the reader confirms the reading and the reading is significantly lower than the historical consumption, the hhu generates a bill with the historical consumption to which the reader does not have access. Once a bill is issued by the hhu, the reader has no authority to cancel it. Instead, they need to report it to the headquarters, and another bill needs to be generated manually after the request is justified. Conversely, if the inserted reading is too high, the hhu provides an alarm, and if the reader confirms, a bill is generated. After bills are generated, a list of customers with bills higher than 100 Jordanian Dinar (J.D.) is produced, and a revenue collector goes to these customers to collect revenues. Considering the size of the property, the revenue collector can also check (based on experience) whether the bill is reasonable. The monthly monetary incentives of the revenue collectors are linked to their collected revenues. Other revenue collection means also exist.

There are several situations in which meter readers are not capable or do not read water meters or record the actual reading in the hhu. These situations are as follows: (i) closed property, (ii) closed meter box, (iii) the reading in the hhu is more than the actual reading in the field, which is common in Zarqa, (iv) unknown location, mainly in new urbanised areas, (v) wet or unclear glass of meter, (vi) meter cannot be reached, (vii) a woman is at home alone, (viii) malfunctioning (stopped) or stolen meters, (ix) insincere or corrupt reader, and (x) work overload for the reader. In these cases, the meter reader estimates the consumption of the customer using the hhu. There are two possibilities for the estimated consumption, either slight underestimations or overestimations.

If a reading is underestimated slightly by mistake, its value disappears in the following cycle and there is no effect on the annual balance. If the reading is underestimated significantly, the hhu generates a bill using historical consumption. However, the reading can also be underestimated significantly and still be more than the historical consumption. This case is likely to be rare because as many as 40%[12] of customers consume less than 18 m^3 per quarter and are charged for this minimum amount. Therefore, there is no benefit for a customer or the reader to decrease the reading for small consumers (40%). Medium consumers are probably not poor because they use more water at a higher tariff block. Reducing the actual consumption of medium customers by 5 m^3 can result in saving only 1 J.D. for both water and wastewater services. The tariff of these blocks is not high enough to make faulty readings profitable (0.185 J.D./m^3). Therefore, consumption under-registration due to the corruption of readers is not likely for medium consumers. For large consumers, the water bill is relatively cheap compared to electricity and other costs (living or production), and their consumption is dynamic and cannot be estimated easily. A corrupt reader does not usually bear such a risk because the data of the hhu are analysed by the staff of the utility and the hhu can possibly generate a bill with historical consumption. Therefore, the underestimation of customer consumption, whether due to corruption or by mistake, is not likely to occur, because of the use and the program of the hhu.

Conversely, overestimation of the consumption is common for the Zarqa water utility. In order to estimate the amount of consumption overestimation in Zarqa, different samples and studies were conducted or collected and then analysed. Before 2015, approximately 8% of the meter readings of customers were overestimated, 50% of which had stopped meters[13]. After 2015, overestimated meters accounted for 16.7% of the total customers in Zarqa and Rosaifah[14]. Among the overestimated meters (16.7% of customers), 4.1% were slightly overestimated and therefore corrected automatically in the following cycles and

---

[12] In some billing cycles; the annual average is 25%.

[13] Based on a sample of 2,767 customers in zone number 7. Source of data: Zallom, S. (2014). "Analysis of bills' complaints and their causes in Zarqa water supply system. Zarqa water utility, Jordan."

[14] Based on analysing records of 148,999 customers. Another sample of 13,441 customers in three different zones has another figure (11%).

12.6% were overestimated significantly to the level that the lag between the actual reading of the meter and the registered reading in the billing system was too large to be corrected unless a new meter was installed[15]. A certain proportion of the 12.6% of customers recognised the increase of their consumption. Of the 12.6% of customers, approximately 2.2%[16] of these customers complained to the utility and asked for a re-check of their bills. Then, their bills were either adjusted or confirmed. The rest either ignored the increase or were unsatisfied with the service of the utility, which contributes to the increase of unauthorised consumption in the network.

To estimate the volume of overestimated consumption in Zarqa, the rate of differences between the actual reading and the billed reading is required. The interviewed staff were too conservative to state, estimate, or provide access to the relevant data. Therefore, different scenario rates were assumed (Table A.3), and the volume of overestimated consumption was estimated to be 2.27% of the SIV. All other errors related to the process of reading customer meters are considered in the 2.27% or corrected in the billing process. In addition, of the 150 readings conducted every day by each meter reader, one reading per week is exposed to an error such as hhu bugs, data transmitting mistakes to the billing system, or different customer serial numbers. These types of errors are discovered and processed during the billing process.

Table A.3. Over-billing due to overestimation of meter readings

Rate of overestimation* (m^3)	Overestimated customers (%)	No. of customers --	Volume of overestimation (m^3)	% of SIV (%)
5	16.7	151,777	506,935	0.76
10	16.7	151,777	1,013,870	1.51
15	16.7	151,777	1,520,806	2.27
20	16.7	151,777	2,027,741	3.03
25	16.7	151,777	2,534,676	3.79
			Average	2.27

* per bill; quarterly

A1.5   BILLING ERRORS

The billing software of the Zarqa water utility is X7 billing software, developed by Adelior France. The software is outdated, and several of its analytical features are not effective. There is no maintenance contract with Adelior France or knowledge transfer. The current version of the software has many shortcomings: (i) X7 accepts negative readings; (ii) it accepts illogical readings, that is, values of millions; and (iii) it allows the

---

[15] Based on analysing records of a sample of 22,489 customers in six different areas.

[16] 2.2% of the customers, including those in the high-elevation parts of the network, complain about their bills. This is based on the above-mentioned sample of 2,767 customers.

alteration and adjustment of the actual consumption without retaining the original figures. This particular disadvantage is crucial as it affects the volume of the billed consumption and non-revenue water. It also has significant financial consequences on the revenue of the utility in the long term as the adjustment and smoothing of water bills is made by altering the actual consumption. Consequently, the historical consumption can eventually decrease, and this historical consumption is used to generate bills for 16.7% of the customers. If there were a separate field in the software for the original consumption, it would enhance and increase the revenues of the utility. This issue is emphasised in the American Water Works Association Water Audit Manual (AWWA 2009). An example of how alteration of the actual consumption is practised by the Zarqa water utility in response to customer complaints is explained in Table A.4.

Table A.4. Example of a bill adjustment

Quarter	Consumption (m3)	Bill Price J.D.		Consumption (m3)	Water Bill J.D.
--					
Q1	50	25	Adjustment=*	73.3	36.7
Q2	50	25		73.3	36.7
Q3	120	120		73.3	36.7
Total	**220**	**170**		**220**	**110**

* The total consumption is the same, however, the quarterly consumption is less and charged within a lower tariff block

Other issues related to the billing errors in the Zarqa water utility include the following:

- The water network is divided into reading areas, for example, there are three areas for the Rosaifa district. The previous practice was to collect readings from the three areas within one month. The current practice is that each area should be read in one month. The division is now time-based, which results in some bills being issued for different periods for the same customer, e.g., two months for one bill and three months for another bill. The high fluctuations of the total amount in the bill lead to customer dissatisfaction.

- Approximately 95% of the bills are transmitted automatically to the billing system, but 5% are transmitted manually. These are susceptible to random errors.

- According to the billing department in Zarqa, among the 50,000 bills produced every month, approximately 2,500 bills (5%) are generated with billing errors such as (i) illogical readings in millions, (ii) mistaken dates or quarters, (iii) negative consumption, (iv) inaccurate consumption, or (v) other types of errors (e.g., the serial number of the customers' meter does not match that in the billing system). Assuming that the distribution of the 5% of errors among these error types is equal, the proportion of each type of error will be 1%. The billing errors reflected in the annual volume of NRW are the last three types. Therefore, the billing errors due to these errors are -1%, $\pm1\%$, and $\pm1\%$, respectively, for negative consumption, inaccurate consumption, and other errors.

- The number of new customers for whom the billing system was not updated were 3,000 in 2014 and 500 in 2015, with an average of 1,750 customers per year (1.2% of the customers).

- If the consumption is negative, then it is billed with the minimum consumption of 18 m³/quarter.

- According to the billing records of 151,778 customers for the four quarters of 2014 and three quarters of 2015, 35,185 customers (23.2%) consumed less than 18 m³ and 2,987 customers (1.97%) consumed 18 m³ (Table A.5). The average actual consumption of the 35,185 customers was 246,964 m³. These customers are billed with the minimum rate, for a total consumption of 633,330 m³; thus, the volume of over-billing due to the minimum consumption policy is 386,366 m³ per quarter and 1,545,465 m³ per year, which accounts for 2.31% of the SIV.

Table A.5. Breakdown of minimum consumption in Zarqa

Range of Consumption	No. of Customers		% of Total Customers	
(m³/quarter)	--	Accumulated	(%)	Accumulated
0 -- 5	4,657	4,657	3	3
6 -- 10	9,943	14,600	7	10
11 -- 15	14,636	29,236	10	19
16 -- 17	5,949	35,185	4	23
18	2,987	38,172	2	25
Total	38,172		25	

The estimated billing errors that form a part of the NRW volume are listed in Table A.6. The overbilling volume is 2.3% of the SIV, which leads to underestimation of the volume of NRW by the same proportion. If the NRW is increased by this amount, the billing errors in scenario 1 in Table A.6 are −4%. If the NRW is not increased by the overbilling percentage, the billing errors remain at −1.7%. As there might be other billing errors not considered in this analysis, scenario 1, which meets the expectations of the commercial department of the Zarqa water utility, is applied. The final result is that the billing errors in Zarqa are −4% as underbilling errors and 2.3% as overbilling errors, with net errors at −1.7% of the SIV.

Table A.6. Estimated billing errors in Zarqa

Type of Billing Error	(%)	Scenario 1	Scenario 2	Scenario 3
Negative Consumption	-1	-1	-1	-1
Wrong Consumption	±1	-1	1	0
Unregistered New Customers	±1	-1	1	0
Min. Cons. Overbilling	2.3	2.3	2.3	2.3
Other	±1	-1	1	0
Total		-1.7	4.3	1.3

A1.6 COMPONENTS OF APPARENT LOSSES

Based on the previous analysis, the errors because of billing are −4% of the SIV, the data handling errors in the form of overbilling are +2.27%, the overbilling due to the minimum consumption policy is +2.31%, and the meter inaccuracy errors are −3.84%. The proportion of total apparent losses is −26.4% of the SIV, so the volume of unauthorised consumption can be calculated. Table A.7 and Figure A.3 show the estimated components of apparent losses in the Zarqa water network.

Table A.7. Volume of subcomponents of apparent losses in Zarqa (2014)

Subcomponent	m^3/year
Meter inaccuracies	-2,570,993
Data handling errors indicate overestimation*	1,520,806
Minimum consumption overbilling	1,545,464
Billing errors	-2,678,043
Unauthorised consumption	-15,492,318
Total apparent losses	-17,675,085

* Due to the program of the hhu, the data handling errors are overestmation

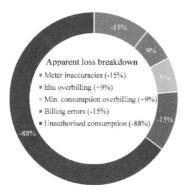

Figure A.3. Components of apparent losses in the Zarqa water network

227

LIST OF ACRONYMS

24/7	Continuous supply
AL	Apparent loss (or commercial loss)
ALC	Active leakage control using leak detection surveys
ALE	Apparent loss estimation equation
ALI	Apparent loss index
AM	Assets management
AWWA	American Water Works Association
AZP	Average zone point
BABE	Bursts and background estimates methodology
BC	Billed consumption
BMC	Billed metered consumption
BUC	Billed unmetered consumption
CAL	Component analysis of the leakage
CAAL	Current annual apparent losses
CARL	Current annual real losses
CI	Intervention cost
CMIs	Customer meter inaccuracies
CHK_1	Logical check = $M_{1b} - M_4 = M_3$
CV	Variable cost
DCMMS	Maintenance Management System
DHEs	Data handling errors
DMA	District metered area
EIF	Economic intervention frequency based on the leakage detection surveys
ELL	Economic level of leakage
EP	Economic percentage of systems to be surveyed annually
EPA	US environmental protection agency
FAVAD	Fixed and variable area discharge

FV	Float-valve
F_{UC}	Unauthorised consumption factor
GWR	Global Water Resources
IBNET	International Benchmarking Network
ICF	Infrastructure condition factor
I/I	Inflow and infiltration
ILI	Infrastructure leakage index
IWA	International Water Association
KPIs	Key performance indicators
LMIC	Low- and middle-income countries
LNC	Legitimate night consumption
M_{1a}	Top-down water balance assuming the unauthorised consumption to be 0.25% of the SIV
M_{1b}	Top-down water balance assuming the unauthorised consumption to be 10% of the BC
M_2	Water and wastewater balance method
M_3	MNF analysis method
M_4	BABE method
MCM	Million cubic metre
MNF	Minimum night flow
MPE	Maximum permissible error
N_1	Discharge exponent; pressure leakage relationship
N_2	Pressure bursts relationship
N_3	Pressure consumption relationship
NDF	Night–day factor
NRW	Non-revenue water
O&M	Operation and maintenance
PIs	Performance indicators
PM	Pressure management
PRVs	Pressure reducing valves

230

q_{max}	The overloaded flow rate (or q_4)
q_{min}	The minimum flow rate (or q_1)
q_p	The permanent flow rate (or q_3)
q_t	The transitional flow rate (or q_2)
RA	Regression analysis
RAAL	Reference annual apparent losses
RL	Real losses
RR	Rate of rise of the unreported leakage
RT	Response time to repair leaks
RTM	Repair response time minimisation
RSS	The residual sum of squares
SIV	System input volume
T_{avg}	Average supply time
UAC	Unbilled authorised consumption
UARL	Unavoidable annual real losses
UBL	Unavoidable background leakage
UC	Unauthorised consumption
UFW	Unaccounted for water
UMC	Unbilled metered authorised consumption
UUC	Unbilled unmetered authorised consumption
WB	Water balance
WL	Water loss
WRC	Water Research Commission
WRF	Water Research Foundation
w.s.p.	When system is pressurised

LIST OF TABLES

LIST OF FIGURES

ABOUT THE AUTHOR

Taha AL-Washali worked for 10+ years as a water professional in the Yemeni water sector. He received his bachelor's degree in civil engineering from Sana'a University, Yemen, and his Master of Science (Excellent) in water resources management from Cologne University, Germany. Currently, he is a program manager at MeaMeta Research. He previously worked in various water projects for governmental and international agencies, including GIZ, KFW, USAID, WHO, and OXFAM. He also contributed to several policies, strategies, and key reports in the Yemeni water sector. Before beginning his doctorate study, he worked as a director general of the projects department at the national water supply and sanitation authority in Yemen. Since then, his research interests have focused on water supply efficiency and water loss management, with particular attention to the needs of developing countries.

Journal publications

AL-Washali, T., Mahardani, M., Sharma, S., Arregui, F., and Kennedy, M. "Impact of Float-Valves on Customer Meter Performance under Intermittent and Continuous Supply Conditions." Resources, Conservation and Recycling, 163, 105091, 2020.

Al-Washali, T., Elkhider, M., Sharma, S., and Kennedy, M. "A Review of Non-Revenue Water Assessment Software Tools." WIREs Water, 7:e1413(02), 2020.

Al-Washali, T., Sharma, S., Lupoja, R., Al-Nozaily, F., Haidera, M., and Kennedy, M. "Assessment of Water Losses in Distribution Networks: Methods, Applications, Uncertainties, and Implications in Intermittent Supply." Resources, Conservation and Recycling, 152(1), 104515, 2020.

AL-Washali, T., Sharma, S., AL-Nozaily, F., Haidera, M., and Kennedy, M. "Monitoring the Non-Revenue Water Performance in Intermittent Supply." Water, 11(6), 1220, 2019.

AL-Washali, T. M., Sharma, S. K., Kennedy, M. D., AL-Nozaily, F., and Mansour, H. "Modelling the Leakage Rate and Reduction Using Minimum Night Flow Analysis in an Intermittent Supply System." Water, 11(1), 48, 2019.

AL-Washali, T. M., Sharma, S. K., and Kennedy, M. D. "Alternative Method for Nonrevenue Water Component Assessment." Journal of Water Resources Planning and Management, 144(5), 04018017, 2018.

AL-Washali, T., Sharma, S., and Kennedy, M. "Methods of Assessment of Water Losses in Water Supply Systems: a Review." Water Resources Management, 30(14), 4985-5001, 2016.

Journal articles submitted/ in preparation

Al-Washali, T., Sharma, S., Mphahlele, T., and Kennedy, M. "Methods of Assessment of Unauthorised Consumption in Water Distribution Networks." Submitted to Water and Environment Journal.

Al-Washali, T., Thilakarathne, W., Sharma, S., and & Kennedy, M. "Artificial Neural Networks for the Detection and Prediction of Unauthorised Water Consumption: Potential Analysis and Future Outlook." In preparation.

Al-Washali, T., Sharma S., and Kennedy, M. "Re-Thinking the Over-billing Practice in Water Distribution Systems: the Case of Zarqa Water Network, Jordan." In preparation.

Conference proceedings

AL-Washali, T., Mahardani, M., Sharma, S., Arregui, F., and Kennedy, M.: Analysing Customer Meter Performance under Intermittency and Continuity Conditions. IWA Water Loss Virtual Conference, Shenzhen, China, 9-11 November 2020.

AL-Washali, T., Sharma, S., Kennedy, M., AL-Nozaily F, and Haidera M.: Normalizing Water Loss Level in Intermittent Supplies. IWA 1st Intermittent Water Supply Conference, Kampala, Uganda, 7-9 April 2019.

AL-Washali, T., Sharma, S., Kennedy, M., AL-Nozaily F, and Haidera M.: Analysis of the Potential of Leakage Reduction Scenarios: Are They Dependent on Each Other? IWA Water Efficient 2019, Manila, The Philippines, 13-16 January 2019.

AL-Washali, T., Sharma, S., Kennedy, M., Lupoja R. P., Shawagfeh Z., AL-Nozaily F, Haidera M., AL-Koli H. S., and AL-Shaieb R.: Comparative Analysis of Non-Revenue Water Assessment Methods in Three Cases in Developing Countries. IWA Water Loss 2018, Cape Town, South Africa, 7-9 May 2018.

AL-Washali, T., Sharma, S., Kennedy, M., AL-Nozaily F, and Haidera M.: Analysing Apparent Losses in Zarqa Water Network, Jordan. Poster Presentation. IWA Water Loss 2018, Cape Town, South Africa, 7-9 May 2018.

AL-Washali, T., Sharma, S., Kennedy, M., AL-Nozaily F, and Haidera M.: Water Loss Assessment and Monitoring in Intermittent Supplies. International Conference for Sustainable Development of Water and Environment, ChenYang, China, 18-19 January 2018.

AL-Washali, T., Sharma, S. & Kennedy, M.: Does A Lower Non-Revenue Water Level Always Mean Better Performance? IWA-Pi 2017 Benchmarking Conference: Specialist Conference on Benchmarking and Performance Assessment, Vienna, Austria, 15-17 May 2017.

AL-Washali, T., Sharma, S. Kennedy, M. & Shwagfeh, Z.: Economic Modelling for Prioritizing Leakage Reduction Measures in AL-Zarqa, Jordan. 4th Arab Water Week, Dead sea, Jordan, 18–23 February 2017.

AL-Washali, T., Sharma, S., and Kennedy, M.: Water Loss Assessment in Developing Countries. IHE-Delft PhD Symposium 2017: Climate Extremes and Water Management Challenges, Delft, The Netherlands, 2-3 October 2017.

AL-Washali, T., Sharma, S., and Kennedy, M.: Water Loss Monitoring. German Arab Master Programs Conference 2017, Berlin, Germany, 2-3 November 2017.

AL-Washali, T.M., Sharma, S., and Kennedy, M.D.: Non-Revenue Water Evaluation in a Complicated Context: The Case of Sana'a Water Supply System. Water Loss 2016 Conference, Bangloure, India, 31 January – 3 February 2016.

Netherlands Research School for the
Socio-Economic and Natural Sciences of the Environment

D I P L O M A

for specialised PhD training

The Netherlands research school for the
Socio-Economic and Natural Sciences of the Environment
(SENSE) declares that

Taha M. AL-Washali

born on 21 June 1983 in Sana'a, Yemen

has successfully fulfilled all requirements of the
educational PhD programme of SENSE.

Delft, 21 December 2020

Chair of the SENSE board

Prof. dr. Martin Wassen

The SENSE Director

Prof. Philipp Pattberg

The SENSE Research School has been accredited by the Royal Netherlands Academy of Arts and Sciences (KNAW)

K O N I N K L I J K E   N E D E R L A N D S E
A K A D E M I E   V A N   W E T E N S C H A P P E N

The SENSE Research School declares that Taha M. AL-Washali has successfully fulfilled all requirements of the educational PhD programme of SENSE with a work load of 60.0 EC, including the following activities:

SENSE PhD Courses

o   Environmental research in context (2015)
o   Research in context activity: Developing and implementing a training course on 'Research Poster design' (2016)

Selection of Other PhD and Advanced MSc Courses

o   Advanced Water Transport and Distribution, IHE-Delft (2019)
o   Data Visualisation, IHE-Delft (2018)
o   Asset Management of Water Systems, IHE-Delft (2018)
o   Resilience, Emergency Preparedness, and Water Safety Planning in Urban Water Systems, Sana'a University, Yemen (2020)
o   Water Meter Management (Field Training), WaterNet (2019)
o   Deep Learning A-Z: Hands-On Artificial Neural Networks, Udemy (2019)
o   Statistical Analysis Using SPSS, International College, Yemen (2016)

Selection of Management and Didactic Skills Training

o   Supervising five MSc students with theses, IHE-Delft (2016-2020)
o   Organizing several national workshops and training courses at Sana'a University and Oxfam for practitioners, MSc, PhD and lecturers (2016-2020)
o   Coordinating a national award: Competition for Good Practices in the Provision of Urban Water Services in Time of Severe Crisis, Yemen (2020)
o   Member of PhD Association Board, IHE Delft (2019-2020)

Selection of Oral Presentations

o   *Economic Modelling for Prioritizing Leakage Reduction Measures in Zarqa, Jordan*. 4th Arab Water Week, 19-23 Mar 2017, Dead Sea, Jordan
o   *Does A Lower Non-Revenue Water Level Always Mean Better Performance?* IWA Benchmarking Conference 2017, 15-17 May 2017, Vienna, Austria
o   *Comparative Analysis of Non-Revenue Water Assessment Methods in Three Cases in Developing Countries.* IWA Water Loss 2018 , 7-9 May 2018, Cape Town, South Africa
o   *Analysis of the Potential of Leakage Reduction Scenarios: Are They Dependent on Each Other?* IWA Water Efficient 2019, 13-16 January 2019, Manila, Philippines
o   *Normalizing Water Loss Level in Intermittent Supplies.* IWA 1st Intermittent Supply Conference, 7-9 April 2019, Kampala, Uganda

SENSE coordinator PhD education

Dr. ir. Peter Vermeulen